T0233727

SpringerBriefs in Philosophy

More information about this series at http://www.springer.com/series/10082

Dingmar van Eck

The Philosophy of Science and Engineering Design

 Springer

Dingmar van Eck
Centre for Logic and Philosophy of Science
Ghent University
Ghent
Belgium

ISSN 2211-4548   ISSN 2211-4556   (electronic)
SpringerBriefs in Philosophy
ISBN 978-3-319-35154-4   ISBN 978-3-319-35155-1   (eBook)
DOI 10.1007/978-3-319-35155-1

Library of Congress Control Number: 2016947386

Printed on acid-free paper

This Springer imprint is published by Springer Nature
The registered company is Springer International Publishing AG Switzerland

Contents

Introduction

Conceptual interactions between philosophy of science and philosophy of engineering (design) are few and far in between. This might be due to several reasons: Most philosophers of engineering (design) seem to think that science and design are two relevantly different kinds of intellectual endeavors (Simon 1969), the philosophy of engineering (design) is still in its 'infancy,' i.e., a young field still in the business of exploring and defining its research agenda (Galle 1999), and until recently engineering has, a few exceptions aside, been ignored by philosophers of science (Calcott 2014; Calcott et al. 2015; van Eck 2015; Braillard 2015). Despite the fact that, for instance, in the case of engineering and biology, researchers from both fields have been stressing (the importance of) conceptual ties for more than a decade (e.g., Csete and Doyle 2002).

In this book, I aim to demonstrate that this mutual lack of attention is an unwelcome situation, for conceptual exchange has the potential to address key issues in both philosophical fields. In addition to mutual enrichment, such interactions may benefit engineering practice itself. I argued for these claims in a variety of papers published in several philosophy of science and engineering design journals, but the approach I defend has never been presented in full detail and in a systematic way. I do so here. In this book, I argue for these claims and spell out my 'explanationist' approach in terms of a 'conceptual common ground' between philosophy of science and philosophy of engineering (design): the related notions of *function* and *explanation*. Specifically, I deploy notions, concepts, and insights from the philosophical literature on scientific explanation to address (related) key issues in the philosophy of technical artifacts and technical functions, and the philosophy of engineering (design). These issues in particular concern the explanatory value of function ascriptions in engineering design and philosophy of technical functions (Chap. 1), and the role and goodness of design and explanatory representations in engineering design and philosophy thereof (Chaps. 3 and 4). These are all pressing and unsolved issues. In advancing these analyses, I also dissolve an alleged key problem in the philosophy of design (Chap. 3)—the notorious 'problem of the absent artifact'—and elaborate means for the testing of design methods (Chap. 4), which benefits engineering practice as well.

Vice versa, I show that scrutiny of engineering practices leads to extension and refinement of models of explanation as discussed in the philosophy of scientific explanation (Chap. 2). I discuss how the mechanistic framework on explanation needs to be extended to capture explanatory practices in engineering, and at the interface of (control) engineering and (systems) biology, in well-informed fashion. Notions of technical function loom large in these analyses. Moreover, these cases serve to illustrate what is required of good mechanistic explanations in different explanation-seeking contexts. The structure of mechanistic explanation in particular fields, in casu engineering science, and assessments of the explanatory power or strength of mechanistic explanations are also important and ongoing topics of analysis in philosophy of science.

As can be gleaned from the above description, this book is meant to serve multiple aims and audiences. Its guiding motivation is that the mutual neglect between philosophy of science and philosophy of engineering (design) is unfounded. Philosophers of engineering design as well as engineering design researchers can benefit from the conceptual toolkit that philosophy of science has to offer. Key issues can be addressed by deploying this toolkit, as exemplified by the fruitfulness of the 'explanationist' approach elaborated in this book. The other way around, philosophy of science can make headway on key issues such as the structure of mechanistic explanation and their explanatory power by taking engineering practices (more) seriously.

As such I hope that this book will be useful to professional/senior philosophers working in philosophy of science and philosophy of engineering (design). It also makes for a useful introductory guide to advanced M.A. and Ph.D. students interested in technical function theories and explanation in engineering science. Lastly, design researchers may benefit from the research on the testing of design methods. The structure of this book reflects these aspirations: Each chapter is self-contained, such that it can be studied in its own right, and does not require knowledge of other chapters.

Although the book is structured such that each chapter is thematically self-contained, the chapters are of course tightly conceptually interwoven. Given the book's focus on technical function and explanation, it starts by assessing in Chap. 1 in which contexts function ascriptions are explanatorily relevant. In Chap. 2, we continue this analysis and also have a closer look at the structure of explanations in which technical functions figure. As we will see, function descriptions are part and parcel of both explanatory representations of the workings of extant technical systems and of design representations of to-be-built ones. We then proceed to assess the role and goodness of these design and explanatory representations in designing in Chaps. 3 and 4, respectively. These latter two chapters thereby also address the issue of the testing of design methods.

References

Braillard, P. A. (2015). Prospects and limits of explaining biological systems in engineering terms. In P. A. Braillard & C. Malaterre (Eds.), *Explanation in biology* (pp. 319–344). Springer.

Calcott, B. (2014). Engineering and evolvability. *Biology and Philosophy, 29,* 293–313.

Calcott, B., Levy, A., Siegal, M. L., Soyer, O. S., & Wagner, A. (2015). Engineering and biology: Counsel for a continued relationship. *Biological Theory, 10,* 50–59.

Csete, M. E., & Doyle, J. C. (2002). Reverse engineering of biological complexity. *Science, 295,* 1664–1669.

Galle, P. (1999). Design as intentional action: A conceptual analysis. *Design Studies, 20,* 57–81.

Simon, H. A. (1969). *The sciences of the artificial.* Cambridge, MA: MIT press.

van Eck, D. (2015). Mechanistic explanation in engineering science. *European Journal for Philosophy of Science, 5*(3), 349–375.

Chapter 1
Assessing the Explanatory Relevance
of Ascriptions of Technical Functions

Abstract In this chapter we assess the explanatory utility of ascriptions of technical functions by considering two explanation-seeking contexts that often figure in the philosophical literature on functions (and explanations). Applied to the technical domain, these are: (i) why was artifact x produced?, and (ii) why does artifact x not have the expected capacity to ϕ? We argue that function ascriptions are explanatorily irrelevant for the first explanation-seeking question, and are explanatorily relevant for the second one. We argue these points in terms of the desideratum that explanations should only list difference making factors.

Keywords Technical function · Explanatory relevance · Difference making

1.1 Introduction

In this chapter we address the explanatory traction of technical function ascriptions. Analysis of technical artifacts and technical functions has proven to be an intricate and rewarding topic of inquiry. Analyses developed in the past ten to fifteen years have shown the initial mainstream assumption that analysis of technical artifacts and technical functions was a rather trivial task for philosophy, and technical functions could easily—and in passing—be accounted for by theories of biological functions, to be untenable (e.g., Preston 1998; Vermaas and Houkes 2003). The phenomenology of technical artifacts and technical functions presents intricacies that are not well accounted for by theories of biological functions. There are now a number of separate analyses focusing on the technical domain, addressing issues such as theories of technical functions (Vermaas 2006; Houkes and Vermaas 2010), mechanistic artifact explanation (De Ridder 2006; De Winter 2011), the epistemology (Houkes 2006) and ontology of technical artifacts (Houkes and Meijers 2006), and comparisons of the dual—intentional and structural—'nature' framework of technical artifacts vis-à-vis collectivist frameworks (Houkes et al. 2011). In this chapter we focus on theories of technical functions.

Although sophisticated theories of technical functions have been advanced in recent years, we argue that something vitally important is missing in current theorizing

© The Author(s) 2016
D. van Eck, *The Philosophy of Science and Engineering Design*,
SpringerBriefs in Philosophy, DOI 10.1007/978-3-319-35155-1_1

about technical artifacts and technical functions, to wit: careful reflection on the *explanatory relevance* of technical function ascriptions. When and why are function ascriptions explanatorily relevant? Whereas the philosophy of biology has made headway on the issue (cf. Wouters 2003), this is not the case for the philosophy of technology. Current philosophical theories of technical functions are mainly concerned with specifying conditions under which agents are *justified* in ascribing functions to technical artifacts (and their components and processes). Yet, assessing the precise *explanatory relevance* of such function ascriptions is, by and large, a neglected topic in the philosophy of technical artifacts and technical functions (cf. Preston 1998; Kroes 2003; Vermaas and Houkes 2003; Krohs 2009; Houkes and Vermaas 2010; van Eck and Weber 2014). The primary aim has been to develop normative accounts for *justifiable function ascription*, rather than making utility assessments of function ascriptions. We address this latter issue in this chapter, using concepts and insights from the philosophical literature on scientific explanation.

We assess the explanatory utility of ascriptions of technical functions by considering two explanation-seeking contexts that often figure in the philosophical literature on functions (and explanations). Applied to the technical domain, these are:

(i)   why was artifact x produced?
(ii) why does artifact x not have the expected capacity to ϕ?

(In Chap. 2 on mechanistic explanation in engineering science, the question (iii) 'How does artifact x realize its capacity to ϕ?' will be dealt with, and in Chap. 3 the predictive value of function ascriptions will be considered.).

In addressing the first question we use the "ICE" theory of technical functions, in which elements from Intentionalist, Causal role, and Evolutionist theories of function are incorporated, as an instrument to assess the relevance of functions ascriptions. We argue that on the basis of the ICE theory, two *parallel explanations* can be constructed for the first explanation-seeking question, a *functional* one that incorporates function ascriptions and a *teleological* one that does not. We argue that, in this explanatory context, the teleological explanation is superior to the functional explanation. The functional explanation black-boxes relevant difference making properties with respect to occurrence of the phenomenon to be explained that are included in the teleological one. This result is not specific to the use of the ICE theory. We argue that when using the alternative function theories of Preston (1998) and Krohs (2009) in this explanation seeking context, function ascriptions also turn out explanatorily irrelevant for the first explanation-seeking question. We therefore conclude that in this context, function ascriptions are—at best—merely heuristically useful in the sense of guiding the construction of adequate explanations, which do not include function ascriptions.

Our analysis of the second explanation-seeking context of explaining artifact malfunction has a different result. By considering an engineering methodology for the analysis of artifact malfunction, developed by Price (1998) and Bell et al. (2007), we show that function ascriptions are useful for black-boxing irrelevant causal details and thereby for focusing on relevant difference making properties with respect to explaining artifact malfunction. In this context, function ascription is required to construct adequate explanations.

In arguing these points we employ a key desideratum from several philosophical accounts of explanation (Woodward 2003; Strevens 2004; Couch 2011; cf. Weisberg 2007) according to which those, and only those, factors that make a difference to (occurrence of) a phenomenon to be explained should be referred to in an explanation. Here, of course, we see a relevant connection between philosophy of technical functions and philosophy of science, viz. explanation and explanatory relevance considerations.

1.2 Functional Versus Teleological Explanation: Why Was Artifact X Produced?[1]

In this section we employ the ICE theory of technical functions (Houkes and Vermaas 2010) as a conceptual instrument to assess the explanatory utility of function ascriptions with respect to the explanation-seeking question:

(i) why was artifact x produced?

We choose to focus in-depth on the ICE theory in our analysis since it, in our view, provides the most sophisticated theory on technical functions, and provides the richest conceptual apparatus to address this question. It invokes more relevant difference-making factors when compared with alternative function theories. Yet, the results we present in this section are not conditional on use of the ICE theory but can be generalized. After our assessment in terms of the ICE theory, we indicate how the alternative function theories of Preston (1998) and Krohs (2009) fare with respect to the above explanation seeking question. As in the case of the ICE theory, also on these alternative theories, function ascriptions turn out heuristic.

1.2.1 The ICE Theory of Technical Functions

The book *Technical functions: on the use and design of artefacts*: *on the use and design of artefacts* (Houkes and Vermaas 2010) contains the most elaborate statement of the ICE theory of technical functions. The normative, rather than descriptive, perspective on justifiable function ascriptions is flagged explicitly in it:

> This choice means that we approach both artefacts and the actions in which they play a role largely from a *normative* rather than a descriptive perspective. We do not offer a theory about how people actually use or design artefacts, or how they in fact describe them in functional terms; instead we seek to provide a framework for evaluating some aspects of

---

[1]This section draws on (van Eck and Weber 2014).

these activities, and we theorise about rational and proper artefact use, and about justifiable function ascriptions. (p. 4)

Houkes and Vermaas (2010) elaborate their ICE-theory by combining insights from three function theories for technical artifacts: the intentional (I) theory (Neander 1991; Bigelow and Pargetter 1987; McLaughlin 2001; Searle 1995), the causal-role (C) theory (Cummins 1975) and the evolutionist (E) theory (Millikan 1989).[2] Function ascriptions to artifacts are analyzed against the background of artifact use and design. The use of an artifact is viewed as the carrying out of a use plan for the artifact. Design is seen as primarily the development of new use plans for artifacts. Another relevant feature is that the theory is agent-oriented rather than property-oriented: the ICE theory takes the form of a theory of *justifiable function ascriptions* by human agents rather than a theory that identifies functions as properties of artifacts.

The core of the theory comprises two definitions of justifiable functions ascriptions (one for designers or justifiers, one for passive users; see 2010, pp. 88–89). These definitions can be merged into a single definition. At the EPSA 2011 symposium in which the book was discussed, Houkes and Vermaas proposed the following general definition, which does not distinguish between the two types of agents:

An agent a justifiably ascribes the physicochemical capacity to ϕ as a function to an item x, relative to a use plan up for x and relative to an account A, if:

I. a believes that x has the capacity to ϕ;
    a believes that up leads to its goals due to, in part, x's capacity to ϕ;
C. a can on the basis of A justify these beliefs; and
E. a communicated up and testified these beliefs to other agents, or a received up and testimony that the designer d has these beliefs.

We will develop our analysis in terms of this definition. As can be seen, the ICE theory is a *normative* theory about *justifiable function ascription*: it concerns when function ascriptions are justified and how they have to be justified.

Although the question *why and under which conditions function ascriptions are explanatorily useful* is—as in other theories of technical function—not explicitly addressed, the ICE theory can be invoked to address this issue. We do so here with respect to the following question:

(i) why was artifact x produced?

1.2.2 Heuristics of Technical Function Ascriptions

We argue that by applying the ICE theory to answer the question why an artifact x was produced, two parallel explanations can be constructed: a *functional* one and a, what

---

[2]Neander's (1991) theory counts as an evolutionist one in the context of biology. Applied to technology, it becomes an intentionalist one (Houkes and Vermaas 2010).

we may call, *teleological* one. Whereas the former, by definition, contains function ascriptions, the latter does not. The question, now, is, which explanation is to be preferred? We address this question in terms of the notion, emphasized in several philosophical accounts of explanation, that those, and only those, factors that make a difference to whether or not a phenomenon to be explained occurs should be specified in an explanation (Strevens 2004, 2008; Couch 2011; cf. Weisberg 2007).[3] Applying this constraint or desideratum has substantive implications: in the explanation-seeking context under consideration, function ascription and functional explanation have a mere heuristic role and, we argue, teleological explanation is to be preferred.

Case 1: backward looking explanation

The first type of cases we consider are questions of the following form:

1. Why was artifact x produced?

Functional explanations, couched in terms of the ICE theory, that we give to answer such questions have the following format:

2. Artifact x was produced because there was a designer d who justifiably ascribed the physicochemical capacity ϕ as a function to x.[4,5]

---

[3]Note that this *desideratum* is different from the theory or model constraint of 'simplicity'. When endorsing 'simplicity' a theorist or modeler may intentionally exclude reference to factors that make a difference to whether or not a phenomenon occurs. The constraint which we endorse here, requires that an agent should strive for describing all the factors that make a difference to whether or not a phenomenon occurs. Whether an agent succeeds in doing so is, of course, a different matter. Weisberg (2007) labels this constraint an "1-causal" representational ideal, and distinguishes it from the representational ideals of "simplicity" and "completeness". The latter requires that an explanation should specify both difference making properties with respect to whether or not a phenomenon occurs, as well as the "higher order causal factors" that affect the precise manner in which the phenomenon occurs (cf. Weisberg 2007, p. 651).

[4]An astute reader may point out that (justified) function ascription could have played no role in answering the first explanation-seeking question since there was no physical artifact yet to which a designer could have ascribed a function to. Agreed, yet our answer is in keeping with the ICE theory: "The historical perspective required to ascribe ICE functions may be limited to the design process; it need not extend to earlier generations of artefacts. *An artefact can therefore straightaway be ascribed the capacity for which designers selected it, even if the artefact is a completely novel one* (the case of the first nuclear plant)" (Houkes and Vermaas 2010, p. 93) (our italics). In other words, the answer accords with the ICE theory. To be sure, we here take function ascriptions as answers to the explanation-seeking question under consideration to be 'proper' function ascriptions. Proper function ascriptions are discussed by Houkes and Vermaas (2010) against the backdrop of what they call 'proper use plans'.

[5]An astute reader may also point out that regarding production, belief initially is sufficient and justified belief only becomes relevant in continuation of the production process. Again, agreed. However, justified belief is central to the ICE theory, both in the ascriptions of functions to technical artifacts, and in accommodating central desiderata put forward in the function literature, such as the proper-accidental function distinction, function ascription in innovative contexts, and the handling of malfunction statements. The underlying reason is that the ICE theory is a "normative rather than a descriptive perspective" on "justifiable function ascriptions" (Houkes and Vermaas 2010, p. 4). Given this perspective, the requirement of justified belief for explaining the

Let us consider an example:

3. Why was the computer mouse produced?

A possible answer is:

4. The computer mouse was produced because there was a designer d who justifiably ascribed the capacity to indicate X–Y positions on computer screens as a function to the computer mouse.

Another possible answer that can be constructed in terms of the key concepts invoked in the ICE theory, is the following non-functional one:

5. The computer mouse was produced because there was a designer d who had a use plan up for it and an account A. d believed (i) that the computer mouse has the capacity to indicate X–Y positions on the computer screens, (ii) that up leads to its goals due to, in part, this capacity. d could on the basis of A justify these beliefs. d communicated up and testified these beliefs to other agents.

So we have here two explanatory formats: a functional explanation (2, exemplified in 4) and a teleological explanation (5, with some details filled in). Now, the latter more elaborate explanatory format naturally leads to several follow-up questions: who was the designer d? What was the use plan s/he had in mind? What was the goal? For instance, the goal may have been to facilitate computer use by feeding commands into the CPU without touching the keyboard. To whom were the beliefs communicated? People to whom the beliefs were communicated may include production managers, financial and marketing managers, and the general manager of the enterprise in which the designer is working.

Given the constraint that an explanation should specify those factors that make a difference to whether or not a phenomenon occurs—here the production of artifact x–, a satisfactory explanation of the fact that the computer mouse was produced should include the details referred to in these additional questions. Information on the designer(s), goal(s), use plan(s), and agents involved in the communication chain(s), is crucial to understand how a given artifact x came to be: a design for a computer mouse without an accompanying use plan for it, nor a specified goal for which it can be employed, and neither a financial and marketing strategy to put the product in the market, simply will not go into production.[6]

(Footnote 5 continued)

production of an artifact is either a bullet one has to bite when adopting the ICE theory, or the ICE theory should be extended to also encompass a descriptive perspective in which 'mere belief' suffices for explaining the production of an artifact. Hence, our use of the term 'justified'.

[6]We focus on those difference making factors that are part of the conceptual framework of the ICE theory, and do not consider other potential difference making factors, such as, say, the choice of materials for the computer mouse. Therefore, our labelling of the notion that explanations should specify difference-making factors as a *desideratum* (cf. note 3). That there are, in the explanatory context under consideration, other difference making factors does not affect the outcome of our comparison of the explanatory superiority of functional vis-à-vis teleological explanations.

Now, the information about the designer can be included without giving up functional talk:

6. The computer mouse was produced because Douglas Engelbart justifiably ascribed the capacity to indicate X–Y positions on computer screens as a function to the computer mouse.

However, the rest of the required information cannot be communicated by means of function talk: from explanation (6) we cannot derive what Engelbart's use plan was, what his account was, to whom he talked, etc. So this explanation has a *heuristic* role: it is a first step towards a more satisfactory explanation. And, importantly, this satisfactory explanation does not employ function talk: *function ascription is removed in order to fill in other, more detailed, information*: his use plan, goals, communication partners, etc.

The point generalizes: explanations that fit in scheme (2) are only a first step, even if we include the name of the designer(s) and the capacity, as we did in (6). The satisfactory explanation requires an implementation of the following scheme:

7. Artifact x was produced because there was a designer d who had a use plan up for it and an account A. d believed (i) that x has the capacity to ϕ, (ii) that up leads to its goals due to, in part, this capacity. d could, on the basis of A, justify these beliefs. d communicated up and testified these beliefs to other agents.

In this teleological scheme, the word 'function' does not occur. Function ascription makes no difference to the phenomenon to be explained. So, in the explanations in which the factors are specified that make a difference with respect to the phenomena to be explained there are no function ascriptions.[7] In other words, *in this context, functional explanations black-box relevant difference making properties with respect to the occurrence of the phenomenon to be explained, which are included in the teleological explanation.*

Importantly, this result is not conditional on use of the ICE theory but generalizes to other theories of technical functions. We make our case in terms of an application of Preston's (1998) pluralist theory of (biological and) technical function and Krohs' (2009) theory of (biological and) technical function. We consider these theories in turn. When applying Preston's (1998) pluralist theory of (biological and) technical function, function ascription also turns out irrelevant with respect to the explanation-seeking question "why was artifact x produced". Preston invokes both the concepts of 'system (or causal role) function' and 'proper function' in the ascription of technical functions to capacities of artifacts. She argues that intended capacities for which artifacts are constructed by designers or inventors initially only have or can be ascribed system/causal role functions (p. 243,

---

[7]Note that the argumentation presented here is not to be confused with conceptual explication of the term 'technical function'. On the ICE account, 'technical function' refers to a physical-chemical capacity. We here invoke the ICE function ascription machinery to construct two parallel explanations.

pp. 249–250). It is only when artifacts continue to be reproduced, that proper functions can be ascribed to those capacities for which the artifacts were reproduced, and this continued production is contingent on successful performance as determined by users, not designers or inventors (pp. 244–245).

Applying Preston's account, a possible answer to the explanation-seeking question "why was artifact x produced" has the following format:

Artifact x was produced because a designer or inventor intended artifact x to perform a certain capacity, to which s/he ascribed a system function.

Now, the last clause 'to which s/he ascribed a system function' adds no explanatorily surplus to the explanation and thus should be removed from it. Explanatorily irrelevant factors have no place in explanations. The fact that a designer or inventor constructed an artifact to perform a certain capacity that s/he desired, suffices. Designers/inventors and desired capacities are the difference making factors here, not the ascription of system functions.

Applying Krohs' (2009) theory leads to the same conclusion that function ascriptions have no added explanatory value in this explanation-seeking context. On Krohs' (2009) account of (biological and) technical function, function is explicated in terms of the causal role concept of function and the notion of 'general design'. General design is defined as the 'type-fixation' of, in the case of technology, components of designed artifacts, i.e., the process by which a configuration/organization of components is brought about. Such processes include construction and assembly plans (pp. 74–75). On this account: "function is the contribution of a type-fixed component to a capacity of a system that is the realization of a design" (p. 79). In the context of artifact designing, a function is 'intended' if a component should make a certain contribution/perform a certain role in order to achieve the goal(s) of a designer (p. 85).

Applying Krohs' account, in the case of components, a possible answer to the explanation-seeking question "why was artifact x produced" has the following format:

Artifact x was produced because a designer intended artifact x to make a certain contribution to a capacity of a system in order to achieve his/her goals.

A possible answer in the case of a system composed of a configuration of components has the following extended format:

Artifact x was produced because a designer intended the components making up the artifact to make certain contributions. The system, in turn, is constructed via type-fixation processes, such as construction and assembly planning.

Again, in both scenarios, the ascription of a function here is irrelevant for explaining artifact production. Designers, goals, construction and assembly plans, and contributions are the difference making factors here. Function ascription adds nothing of explanatory relevance.

In the next section we consider the explanation-seeking context of malfunction explanation. Here, the situation is very different: we argue that the explanatory leverage of function ascriptions precisely consists in black-boxing explanatorily irrelevant details.

1.3 Malfunction Explanation

We have seen that functional explanations—explanations containing one or more function ascriptions in the explanans—are not optimal for explaining why an artifact x was produced: there is a non-functional/teleological alternative that is better. We now address a second explanatory context: diagnostic reasoning. In this context, we argue, the situation is reversed: function ascriptions are explanatorily relevant here and functional explanations provide the most adequate explanations. We make our case in terms of discussing an engineering methodology for malfunction analysis.

A widely adopted desideratum in the literature on technical functions is that function theories should advance a notion of *proper* function that allows malfunctioning. In different accounts, this is done in different ways. On the ICE theory, agents that ascribe functions to capacities of artifacts should be able to justify their beliefs that those artifacts have these capacities on the basis of either experience, testimony, or scientific or technological knowledge (the account *A*). Nevertheless, this measure of support, in principle, leaves open the possibility that an artifact malfunctions, despite the agent's (erroneous) belief that the artifact does have the capacity. Hence, malfunction is accommodated within the ICE theory. Krohs (2009) proceeds in different fashion. Rather than justified yet erroneous belief as in the ICE theory, in Krohs' theory, the notion of type fixation determines standards for the contributions of components which they can fail to achieve. Similarly, in the account of Preston (1998) successful performance as measured by users provides a yardstick to accommodate malfunction. Yet, of course, *the accommodation of malfunctioning artifacts within schemes for the ascription of functions to technical artifacts, is completely different from explaining the occurrence of malfunctioning artifacts*. Notions like 'Justified yet erroneous belief' (ICE theory), 'unsuccessful performance as measured by users' (Preston), and 'not meeting standards for components' contributions' (Krohs) are not difference making factors that explain the occurrence of specific malfunctions. Malfunction explanation requires (contrastive) explanation that isolates the specific fault(s) that cause malfunction(s).

Therefore, we leave theories of technical functions here aside and focus on engineering diagnostic reasoning methods that are aimed to explain occurrences of malfunctions in technical artifacts, and we clarify the structure of the explanatory formats that these methods advance, to wit: *contrastive* functional explanations.

1.3.1 Malfunction Analysis: An Engineering Example

When an artifact does not serve or fulfill a function which we expect it to do, explanation-seeking questions of the following format arise:

Why does artifact x not serve the expected function to ϕ?

For instance: why does my heating device not fulfill the expected function to increase its surface temperature? Or: why does my electric screwdriver fail to drive screws?[8]

These questions are *contrastive*: they contrast the actual situation with an ideal and expected one (cf. Lipton 1993). An explanation of a contrast (e.g., why does the heater fail to increase its surface temperature) picks out those causes that are taken to *make a difference* to the occurrence of the phenomenon to be explained, in this case a particular malfunction (van Eck and Weber 2014). That is, contrastive explanations describe those factors that explain, make a difference to, the contrast drawn in the explanandum 'why malfunction, rather than normal function'. For instance, a damaged component that normally converts electricity-into-torque might be responsible for the electric screwdriver's failure to drive screws.[9]

This brief description of the structure of malfunction explanation signals the need for a tool to highlight and specify those contrastive factors that explain the difference between malfunctioning artifacts and normally functioning ones. Function ascriptions are clearly a useful tool for this task: specifying the difference between normal and impaired function (i.e., explaining what has gone wrong) can be done by claiming that a component or sub mechanism *malfunctions*. For example, the claim that 'the component sub serving electricity-to-torque conversion malfunctions'.

The explanatory utility of function ascriptions can be made more precise by considering two constraints derived from the engineering functional modeling literature on malfunction explanation. These constraints are: (1) the ability to black box irrelevant details and make salient relevant details of technical systems, and (2) the ability to identify contrastive difference makers, i.e., malfunctioning components or sub mechanisms (cf. Sembugamoorthy and Chandrasekaran 1986; Price 1998; Hawkins and Woollons 1998; Bell et al. 2007; van Eck and Weber 2016).

Both constraints are endorsed in the engineering literature on fault analysis, and the first one is also clearly exemplified by our description of the structure of a malfunction explanation. Explanations for specific malfunctions cite contrastive difference makers that explain the contrast between malfunction and normal function, possibly enriched with some further details that enable understanding how

---

[8]Varieties of this general question-format are for instance: 'why does component x function suboptimal?' (cf. Otto and Wood 2001)?; 'why is this unexpected and undesired behavior present?'; 'which malfunction is responsible for the undesired behavior?'; 'which components/module does not work as expected?' (cf. Goel and Chandrasekaran 1989; Bell et al. 2007); 'does the trigger of the function fail and/or its effect?' (Bell et al. 2007).

[9]The explanation might also list some further 'local details' that enable understanding how specific features make a difference to a specific malfunction. For instance, oil leaking into the hot exhaust due to a rupture in the oil reservoir may cause a car to expel thick black smoke. One can imagine that some further details are relevant to understand this malfunction, say, the exhaust function of expelling (normal amounts of) smoke and the carburetor producing sparks, since sparking is a cause of both expulsion of normal and excess amounts of smoke. More on this 'enrichment' of malfunction explanations with specific mechanism details in Chap. 2.

specific features make a difference to a specific malfunction (cf. note 9). Those details that underlie normal function but do not increase an explanation's explanatory traction for a specific malfunction should be left out.[10]

We illustrate these constraints by way of empirical examples derived from an engineering methodology for malfunction analysis of technical artifacts, called Functional Interpretation Language (FIL), developed by Bell et al. (2005, 2007), and asses the utility of function ascription in terms of these constraints. We choose to focus on the FIL methodology since it gives a clear exposition of these constraints.

The FIL methodology was developed and is used in industry for a variety of diagnostic reasoning tasks, in particular Failure Mode and Effect Analysis (FMEA). In short, in FMEA analyses, the effects of a malfunctioning component on the overall behavior of an artifact are analyzed, by comparing the overall behavior of artifacts working correctly with the overall behavior of ones that do not, due to a component failure/malfunction (e.g., Price 1998; Hawkins and Woollons 1998; Bell et al. 2007). In FIL, functions are represented in terms of three elements: the *trigger* of a function, its associated and expected *effect*, and the *purpose* that the function is to fulfill. Triggers describe input states that actuate physical behaviors which result in certain (expected) effects. So triggers are the input conditions for effects, i.e., functions, to be achieved.[11] Purposes describe desired states of affairs in the world that are achieved when a trigger results in an expected effect (Bell et al. 2007, p. 400). For instance, with FIL, the function of a cooking ring of a cooking hob is described in terms of the trigger "switch on", the effect "heat ring", and the purpose "cook on ring" (cf. Fig. 1.1). This description is a summary of some salient features of (manipulating) such artifacts; throwing the switch will, if the system functions properly, result in the heating of the ring(s), which in turn supports the preparation of food.

According to Bell et al. (2007) such trigger and effect representations serve two explanatory ends in malfunction analyses: firstly, they *highlight* relevant behavioral features, i.e., effects, and, simultaneously, provide the means to *ignore* less relevant or irrelevant behavioral features, i.e., physical behaviors underlying these effects, of a given artifact; secondly, they support assessing which components are malfunctioning (pp. 400–401).

For instance, the trigger-effect representation "switch on"-"heat ring" highlights the input condition of a switch being thrown, and the resulting desired effect of heat, yet ignores the structural and behavioral specifics of the switch and ring, as well as the energy conversions—e.g., electricity conversions into thermal energy—that are needed to achieve this effect. Such representations only highlight those features that

---

[10]Malfunction explanations already require various assumptions about the structure of a system, of course: a lot of structural and behavioral knowledge is involved (cf. Goel and Chandrasekaran 1989; Bell et al. 2007). This knowledge serves as backdrop against which to assess which features are explanatorily relevant and thus get referred to in the function descriptions.

[11]Triggers are inputs for main or primary normal functions and provide 'pointers' to possible malfunctions (as will become clear later on). Triggers are thus different from 'control functions' which are intended to counteract unwanted disturbances and unwanted changes in engineering systems (cf. Lind 1994).

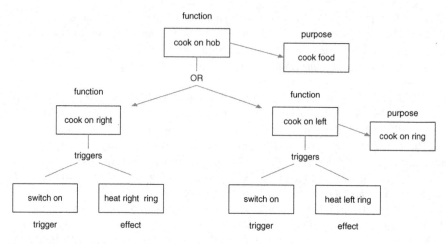

Fig. 1.1 Functional decomposition of a two-ring cooking hob [the example is drawn from Bell et al. (2007); the diagrammatic representation is based on Bell et al. (2005)]

are considered explanatorily relevant to assess malfunctioning systems, and omit reference to physical behaviors/energy conversions by which desired effects are achieved.

There is another way in which the use of trigger-effect descriptions is considered an explanatory asset in highlighting explanatorily relevant features in malfunction explanation: comparing normally functioning technical systems with malfunctioning ones in order to identify malfunctioning components or sub mechanisms (Bell et al. 2007). Trigger-effect descriptions support assessing whether the expected effects in fact obtain, and, if not, which and how components are malfunctioning (Bell et al. 2007). A normally functioning technical system, say the cooking hob, has both a trigger and an effect occurring; a switch is thrown and a ring is heated. Trigger-effect descriptions support analysis of two varieties of malfunction. First, a trigger may occur, but fail to result in the intended effect. Say, the switch is on, yet the ring fails to heat. Second, a trigger may not be occurring, yet the effect is nevertheless present. Say, the switch is not on, but the ring is nevertheless heated (see Bell et al. 2007). Such analyses of the actual states of triggers and effects allows one to focus *on the most likely causes of failure* (Bell et al. 2007). Say, if the switch is on and the ring fail to heat, first likely causes to investigate may be whether the electrical circuitry connected to the ring is damaged. On the other hand, if the switch is not on and the ring is heated, a first likely cause to investigate may be whether the 'on/off' display of the switch is damaged (we here see, as mentioned in note 10, that structural knowledge of a system serves as backdrop for malfunction assessment). The functional decomposition model in Fig. 1.1 is an example of the sort of models used for malfunction explanation.

In such assessments, elaborate details on structural and behavioral specifics of technical artifacts, e.g., all the details of the hob's electrical circuit, are considered unwanted. Functional descriptions pick out only the salient and relevant details for malfunction assessment. (Of course, after a likely cause (or causes) of a particular malfunction has been identified, it may become useful for an analyst to investigate a malfunctioning component or sub mechanism in more detail, say, for repair or redesign purposes. More detailed behavioral models of components and their behaviors are used in FIL for this task, but only after a first round functional analysis of malfunction. Such behavioral analyses may reveal multiple features underlying a component malfunction, say, multiple faults in a cooking hob's electrical circuit).

Thus here we have a case in which function ascriptions and malfunction claims are clearly relevant. Functional descriptions in FIL highlight the salient features of normally functioning artifacts, suppressing reference to irrelevant behavioral and/or structural details (constraint 1), and pinpoint were the specific faults occur in malfunctioning artifacts (constraint 2). Functional descriptions here thus satisfy the two constraints on malfunction explanation which we introduced earlier.

Note that this example clearly contrasts with our first case where a functional explanation couched in terms of the ICE theory leaves out information that is relevant (see explanation scheme 6), and additional details should be included to arrive at a satisfactory explanation (see the complete explanation scheme 7, which does not include function ascriptions).

1.4 Conclusions

In this chapter we disproved the assumption, quite often made in the philosophical literature on functions, that function ascriptions in themselves are explanatorily relevant (e.g. Wright 1973; Millikan 1989; Neander 1991). Wimsatt (1972) and Wouters (2003) already cautioned against the idea that function ascriptions, by definition, provide explanations. Whether or not function ascriptions have explanatory leverage is an issue that requires careful analysis. In this chapter we assessed the explanatory relevance of ascriptions of technical functions, an issue by and large neglected in the literature on technical artifacts and technical functions. We analyzed the relevance of technical function ascriptions in two different explanatory contexts. We argued that whereas function ascriptions serve a mere heuristic role in the context of explaining why artifacts are produced, they play a substantial role in explaining artifact malfunction.

References

Bell, J., Snooke, N., & Price, C. (2005). *Functional decomposition for interpretation of model based simulation. Proceedings of the 19th International Workshop on Qualitative Reasoning, QR-05*, 192–198.

Bell, J., Snooke, N., & Price, C. (2007). A language for functional interpretation of model based simulation. *Advanced Engineering Informatics, 21*, 398–409.

Bigelow, J., & Pargetter, R. (1987). Functions. *Journal of Philosophy, 84*, 181–196.

Couch, M. (2011). Mechanisms and constitutive relevance. *Synthese, 183*, 375–388.

Cummins, R. (1975). Functional analysis. *Journal of Philosophy, 72*, 741–765.

De Ridder, J. (2006). Mechanistic artefact explanation. *Studies in History and Philosophy of Science, 37*, 81–96.

De Winter, J. (2011). A pragmatic account of mechanistic artifact explanation. *Studies in History and Philosophy of Science, 42*(4), 602–609.

Goel. A., & Chandrasekaran, B. (1989). *Functional representation of designs and redesign problem solving. Proceedings Eleventh International Joint Conference on Artificial Intelligence (IJCAI-89)*, Detroit, Michigan, August, 1989: 1388–1394.

Hawkins, P. G., & Woollons, D. J. (1998). Failure modes and effects analysis of complex engineering systems using functional models. *Artificial Intelligence in Engineering, 12*(4), 375–397.

Houkes, W. (2006). Knowledge of artifact functions. *Studies in History and Philosophy of Science, 37*, 102–113.

Houkes, W., & Meijers, A. (2006). The ontology of artefacts: the hard problem. *Studies in History and Philosophy of Science, 37*, 118–131.

Houkes, W., & Vermaas, P. E. (2010). *Technical functions: On the use and design of artefacts.* Dordrecht: Springer.

Houkes, W., Kroes, P., Meijers, A., & Vermaas, P. E. (2011). Dual-Nature and collectivist frameworks for technical artefacts: a constructive comparison. *Studies in History and Philosophy of Science, 42*, 198–2005.

Kroes, P. (2003). Screwdriver philosophy Searle's analysis of technical functions. *Techné, 6*(3), 22–35.

Krohs, U. (2009). Functions as based on a concept of general design. *Synthese, 166*, 69–89.

Lind, M. (1994). Modeling goals and functions of complex industrial plants. *Applied Artificial Intelligence, 8*, 259–283.

Lipton, P. (1993). Making a difference. *Philosophica, 51*, 39–54.

McLaughlin, P. (2001). *What functions explain.* Cambridge: Cambridge University Press.

Millikan, R. (1989). In defense of proper functions. *Philosophy of Science, 56*, 288–302.

Neander, K. (1991). The teleological notion of "function". *Australasian Journal of Philosophy, 69*, 454–468.

Otto, K. N., & Wood, K. L. (2001). *Product design: Techniques in reverse engineering and new product development.* Upper Saddle River NJ: Prentice Hall.

Preston, B. (1998). Why is a wing like a spoon? A pluralist theory of functions. *Journal of Philosophy, 95*, 215–254.

Price, C. J. (1998). Function-directed electrical design analysis. *Artificial Intelligence in Engineering, 12*(4), 445–456.

Searle, J. (1995). *The construction of social reality.* New Haven: Free Press.

Sembugamoorthy, V., & Chandrasekaran, B. (1986). Functional representation of devices and compilation of diagnostic problem-solving systems. In J. Kolodner & C.K. Riesbeck (Eds.), *Experience, memory, and reasoning.* Hillsdale, NJ: Lawrence Erlbaum Associates: 47–53.

Strevens, M. (2004). The causal and unification approaches to explanation unified—causally. *Noûs, 38*(1), 154–176.

Strevens, M. (2008). *Depth: An account of scientific explanation.* Harvard University Press.

van Eck, D., & Weber, E. (2014). Function ascription and explanation: Elaborating an explanatory utility desideratum for ascriptions of technical functions. *Erkenntnis, 79*, 1367–1389.

van Eck, D., & Weber, E. (2016). In defense of co-existing engineering meanings of function. *Artificial Intelligence for Engineering Design, Analysis, and Manufacturing.* Online first, doi:10.1017/S0890060416000172.

Vermaas, P. E. (2006). The physical connection: engineering function ascriptions to technical artefacts and their components. *Studies in History and Philosophy of Science, 37*, 62–75.

Vermaas, P. E., & Houkes, W. (2003). Ascribing functions to technical artefacts: A challenge to etiological accounts of functions. *British Journal for the Philosophy of Science, 54*, 261–289.

Weisberg, M. (2007). Three kinds of idealization. *The Journal of Philosophy, 104*(12), 639–659.
Wimsatt, W. C. (1972). Teleology and the logical structure of function statements. *Studies in History and Philosophy of Science, 3,* 1–80.
Woodward, J. (2003). *Making things happen.* Oxford: Oxford University Press.
Wouters, A. G. (2003). Four notions of biological function. *Studies in History and Philosophy of Biology and Biomedical Science, 34,* 633–668.
Wright, L. (1973). Functions. *Philosophical Review, 82,* 139–168.

Chapter 2
Mechanistic Explanation in Engineering Science

Abstract Explanation already loomed large in Chap. 1 on the explanatory utility of function ascriptions in engineering. In this chapter we take a closer look at the structure of (mechanistic) explanation in engineering. This analysis highlights different meanings that engineers attach to the notion of function, and clarifies the explanatory relevance of this ambiguity, it suggests an extension of the mechanistic program when applied to engineering science and, moreover, contains general lessons on the explanatory power of mechanistic explanations. In explicating the structure of mechanistic explanation, we will also address the question (iii) 'How does artifact x realize its capacity to ϕ?' and the relevance of function ascription in procuring an answer to this question. (we will address this relevance both for type and token-level cases).

Keywords Mechanistic explanation · Function · Engineering · Explanatory power

2.1 Introduction

Use of 'mechanism talk' is ubiquitous in both engineering science (e.g., Chandrasekaran and Josephson 2000; Goel 2013) and philosophical discussions of mechanisms (cf. Levy 2014). Engineered systems, such as pumps, car engines, mouse traps, toilets, soda vending machines, and the like are frequently used in illustrating various aspects of mechanisms and mechanistic explanation. Despite this reference to engineered systems in discussions of mechanisms and mechanistic explanation, focused philosophical analyses of the structure of mechanistic explanations in engineering science are scarce (cf. van Eck 2015a). There is very few philosophical work on engineering mechanisms that does more than (merely) use engineering mechanisms as a loose metaphor, and actually offers sophisticated understanding of what mechanistic explanation looks like in engineering practice. Moreover, although practicing engineers and biologists have been stressing conceptual ties between their disciplines for more than a decade (e.g., Csete and Doyle 2002), this connection has also received scant attention by philosophers, in

© The Author(s) 2016
D. van Eck, *The Philosophy of Science and Engineering Design*,
SpringerBriefs in Philosophy, DOI 10.1007/978-3-319-35155-1_2

particular with respect to the use of engineering principles in the construction of mechanistic explanations in systems biology (cf. Braillard 2015). In this chapter I aim to make headway on both these issues.

In this chapter I give an outline of the structure of mechanistic explanation in engineering science, and organize this discussion around the usage of different meanings of technical function in engineering practice. I show that depending upon explanatory context, engineers use different conceptions of role function, i.e., *behavior* function and *effect* function, to individuate technical mechanisms and to develop mechanistic explanations. I argue that in order to capture this explanatory diversity, and thus to understand mechanistic explanation in engineering science, the mechanistic concept of role function needs to be regimented into these two domain-specific subtypes of role function when applied to the engineering domain. I illustrate this connection between subtypes of role function and explanatory requests in Sect. 2.2 in terms of token and type-level capacity explanations and in terms of malfunction explanations. The general insight that I take these cases to convey is thus that empirically-informed understanding of mechanistic explanation in engineering science requires sensitivity to this distinction in sub types of role function (van Eck 2015a).

In addition, in Sect. 2.3, I briefly discuss connections between (control) engineering and systems biology, focusing on the usage of engineering principles in the construction of mechanistic explanations in systems biology. Systems biology has adopted engineering tools and principles, in particular from control engineering, to model and explain complex biological systems. These tools are often in the service of characterizing the organization of mechanisms in abstract, truncated fashion. I briefly discuss a case of heat shock response in *Escherichia coli* to illustrate the role of engineering principles in mechanistic explanation in systems biology (cf. El-Samad et al. 2005; Braillard 2015). In this case, again, the distinction between the two subtypes of technical role function proves explanatorily relevant.

In Sect. 2.4, I revisit the engineering cases on capacity and malfunction explanation and argue that they give novel, general insights on the explanatory power of mechanistic explanations. I flesh out the distinctions between the explanatory desiderata of 'completeness and specificity' (Craver 2007) and 'abstraction' (Levy and Bechtel 2013) that are stressed in recent discussions on the explanatory power of mechanistic explanations in terms of these cases and argue that, rather than being in competition, as some authors have it, these desiderata are suitable for different explanation-seeking contexts. Furthermore, I argue that both desiderata fall short in the context of malfunction explanation, since they pull in opposite directions there, and elaborate a novel desideratum that can handle this explanatory context better. This desideratum, I argue, is applicable to both engineering and biological contexts of malfunction explanation.

2.2 Mechanistic Explanation in Engineering Science

2.2.1 Mechanistic Explanation: Explanation by Decomposition and (Role) Function Ascription

In this section, we will first have a brief look at the general structure of mechanistic explanation and then apply (and extend) the framework to engineering science. Although there are several accounts of mechanistic explanation on offer in the literature, there is broad consensus on a number of key features:

> All mechanistic explanations begin with (a) the identification of a phenomenon or some phenomena to be explained, (b) proceed by decomposition into the entities and activities relevant to the phenomenon, and (c) give the organization of entities and activities by which they produce the phenomenon. (Illari and Williamson 2012: 123).

Mechanistic explanations thus explain how mechanisms, i.e., organized collections of entities and activities, produce phenomena (Machamer et al. 2000; Glennan 2005; Bechtel and Abrahamsen 2005; Craver 2007). In the literature on explanation in the life sciences, it is now widely recognized that mechanisms play a central role in explaining complex capacities such as digestion, pattern recognition, or the maintenance of circadian rhythms. The idea is that to explain such capacities, one provides a model, or more generally a description/representation, of the mechanism responsible for that capacity.

Role function ascription plays a key role in the (b) decomposition of mechanisms (c) and the elucidation of their organization (Machamer et al. 2000; Craver 2001; Illari and Williamson 2010). As Machamer et al. (2000) write:

> Mechanisms are identified and individuated by the activities, and entities that constitute them, by their start and finish conditions, and by their functional roles. Functions are the roles played by entities and activities in a mechanism. To see an activity as a function is to see it as a component in some mechanism, that is, to see it in a context that is taken to be important, vital, or otherwise significant. (Machamer et al. 2000: 6)

Mechanistic role functions thus refer to activities that make a contribution to the workings of mechanisms of which they are a part, and mechanistic organization is key for the ascription of functions. For instance, in the context of explaining the circulatory system's activity of "delivering goods to tissues", the heart's "pumping blood through the circulatory system" is ascribed a function relative to organizational features such as the availability of blood, and the manner in which veins and arteries are spatially organized (Craver 2001: 64).

There is broad consensus in the literature on mechanistic explanation in the life sciences on the above-mentioned key features of mechanistic explanation, as well as on the importance of (role) function ascription and the functional individuation of mechanisms. And the strong suggestion that one can find in this literature is that the (functional) individuation of mechanisms proceeds in similar fashion in engineering science: frequently, mechanisms of technical artifacts, such as clocks, mousetraps, and car engines, are invoked as metaphors to elucidate features of biological mechanisms (Craver 2001) and features of mechanisms in general (Glennan 2005; Darden 2006; Illari and Williamson 2012). The mechanistic concept of role function, and its utility in the functional individuation of mechanisms, has likewise been explicated in terms of mechanisms of technical artifacts such as car engines (Craver 2001). At the same time however, rigorous analysis of mechanistic explanatory practices in engineering are few and far in between. This invites the question whether the general framework on mechanistic explanation and mechanism individuation, as it is taken to work in the life sciences, can indeed be applied without significant modifications to engineering and able to provide understanding of mechanistic explanation in this domain.

In this chapter I argue that reference to such technical mechanisms is a loose metaphor and must not be understood as proving insight into mechanistic explanation in engineering science per se (cf. van Eck 2015a). In engineering science, technical mechanisms are not functionally individuated in terms of the concept of role function *simpliciter*. Rather, different notions of engineering function, 'behavior function' and 'effect function', are invoked to individuate technical systems and to explain their workings (van Eck 2015a). In order to capture mechanistic explanatory practices in engineering in well-informed fashion, the general perspective on the functional individuation of mechanisms thus needs to be extended to include both senses of engineering (role) function. In the next section I present the conceptual groundwork for this claim by briefly discussing how these varieties of function are used in mechanism individuation and mechanistic explanation in engineering science.

2.2.2 *Function and Functional Decomposition in Engineering*

Function is a key term in engineering (e.g., Chandrasekaran and Josephson 2000). Descriptions of functions figure prominently in, for instance, design methods (Stone and Wood 2000), reverse engineering analyses (Otto and Wood 2001), and diagnostic reasoning methods (Bell et al. 2007).

Despite the centrality of the term, function has no uniform meaning in engineering: different approaches advance different conceptualizations (Erden et al. 2008), and some researchers use the term with more than one meaning simultaneously (Chandrasekaran and Josephson 2000). This ambiguity led to philosophical analysis of the precise meanings of function involved. Vermaas (2009) regimented the spectrum of available function meanings into three '*archetypical*' engineering conceptualizations of function: *behavior function*–function as the desired behavior of a technical artifact; *effect function*–function as the desired effect of behavior of a technical artifact; *purpose function*–function as the purpose for which a technical artifact is designed.[1] In the ensuing discussion, the notions of behavior function and effect function are (most) relevant.

Behavior functions are typically modeled as conversions of flows of materials, energy, and signals, where input flows and output flows in the conversion (are assumed to) match in terms of physical conservation laws (Stone and Wood 2000; Otto and Wood 2001). For instance, the function "loosen/tighten screws" of an electric screwdriver is then represented as a conversion of input flows of "screws" and "electricity" into corresponding output flows of "screws", "torque", "heat", and "noise" (cf. Stone and Wood 2000: 364). Since these descriptions of functions are specified such that input and output flows match in terms of physical conservation laws, they are taken to refer to specific physical behaviors of technical artifacts (Vermaas 2009).

Effect function descriptions refer to only the technologically relevant *effects* of the physical behaviors of technical artifacts: the requirements are dropped that descriptions of these effects meet conservation laws and that matching input and output flows are specified (Vermaas 2009). The function of an electric screwdriver is then described simply as, say, "loosen/tighten screws", leaving it unmentioned what the physical antecedents are of this effect. Behavior function descriptions thus refer to the 'complete' behaviors involved, including features like thermal and acoustic energy flows, whereas effect functions refer to subsets of these behaviors, i.e., desired effects.

Engineering descriptions and explanations of the workings of extant technical artifacts and artifact designs are often constructed by functionally decomposing functions into a number of sub functions. The relationships between functions and sets of their sub functions are often graphically represented in functional decomposition models. Like the concept of function, such models come in a variety of 'archetypical' flavors (van Eck 2011). For our purposes, the relevant ones are *behavior functional decomposition*—a model of an organized set of behavior functions, and *effect functional decomposition*—a model of an organized set of effect functions.

The use of (varieties of) functional decomposition is ubiquitous in engineering science in a variety of tasks, like conceptual engineering design (Stone and Wood 2000), failure analysis (Bell et al. 2007), and reverse engineering and redesign (Otto and Wood 2001). Cases in point are, amongst others, reverse engineering explanations which use elaborate behavior functions and functional decompositions, and malfunction explanations which use less detailed effect functions and functional decompositions.

---

[1]The term 'archetypical' here refers to 'most common'; the three conceptualizations of function are not meant to be exhaustive. For instance, some engineers use 'function' to refer to intentional behaviors of agents (cf. van Eck 2010). In reverse engineering analyses, 'function' refers to actual or expected behavior, without the normative connotation 'desired'.

2.2.3 Reverse Engineering Explanation (and Redesign): Token Level Capacity Explanation

In engineering science, reverse engineering and engineering design go hand in glove (e.g. Otto and Wood 2001; Stone and Wood 2000). Consider Otto and Wood's (2001) reverse engineering and redesign method, in which a reverse engineering phase in which reverse engineering explanations are developed for existing artifacts, precedes and drives a subsequent redesign phase of those artifacts. The goal of the reverse engineering phase is to explain how existing artifacts produce their overall (behavior) functions in terms of underlying mechanisms, i.e., organized components and sub functions (behaviors) by which overall (behavior) functions are produced. That is, reverse engineering—mechanistic—explanations give an answer to the question 'How does a particular artifact x realize its capacity to ϕ?'. These explanations of token level capacities are subsequently used in the redesign phase to identify components that function sub optimally and to either improve them or replace them by better functioning ones.

In the reverse engineering phase, an artifact is first broken down component-by-component, and hypotheses are formulated concerning the functions of those components. In this method, functions are behavior functions and represented by conversions of flows of materials, energy, and signals. After this analysis, a different reverse engineering analysis commences in which components are removed, one at a time, and the effects are assessed of removing single components on the overall functioning of the artifact. Such single component removals are used to detail the functions of the (removed) components further. The idea behind this latter analysis is to compare the results from the first and second reverse engineering analysis in order to gain potentially more nuanced understanding of the functions of the components of the (reverse engineered) artifact. Using these two reverse engineering analyses, a behavior functional decomposition of the artifact is then constructed in which the behavior functions of the components are specified and interconnected by their input and output flows of materials, energy, and signals (Otto and Wood 2001). Such models represent parts of the mechanisms by which technical systems operate, to wit: causally connected behaviors of components. Examples of an overall behavior function and behavior functional decomposition of a reverse engineered electric screwdriver are given in Figs. 2.1 and 2.2, respectively.

In the model in Fig. 2.2, temporally organized and interconnected behaviors are described. Components of artifacts are described in Otto and Wood's method in tables, what in engineering are called 'bills of materials', together with a model, called 'exploded view', of the components composing the artifacts. Taken together, these component and behavior functional decomposition models provide functional individuations and representations of mechanisms of artifacts.

Such (behavior functional decomposition) models are subsequently used to identify sub-optimally functioning components and so drive succeeding redesign phases (Otto and Wood 2001). The focus here is on the reverse engineering explanation-part of the methodology.

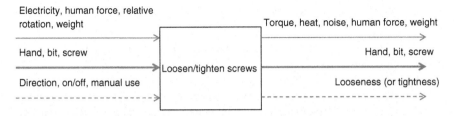

Fig. 2.1 Overall behavior function of an electric power screwdriver. *Thin arrows* represent energy flows; *thick arrows* represent material flows, *dashed arrows* represent signal flows (adapted from Stone and Wood 2000: 363, Fig. 2)

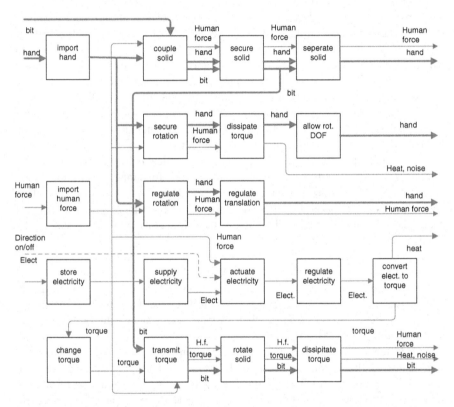

Fig. 2.2 Behavior functional decomposition of an electric power screwdriver. *Thin arrows* represent energy flows; *thick arrows* represent material flows, *dashed arrows* represent signal flows (adapted from Stone and Wood 2000: 364, Fig. 4; cf. Stone et al. 1998, 2000)

In malfunction explanation, this detail in mechanistic models is however not required: engineers take it that less detailed effect functions and functional decompositions there do a better explanatory job (see Chap. 1).

2.2.4 Malfunction Explanation

As we saw in Chap. 1, in malfunction analysis, explanation-seeking questions of the following format arise:

Why does artifact x not serve the expected function to ϕ?

Such questions are *contrastive*: why malfunction, rather than normal function? In the engineering literature, malfunction explanations that answer contrastive questions list different and fewer mechanistic features than reverse engineering explanations which answer questions about normal behavior or function. Such explanations are constructed using effect functions and functional decompositions.

Malfunction explanations in engineering pick out only a few features of mechanisms, i.e., those causal factors—failing components or sub mechanisms—that are taken to make a difference to the occurrence of a specific malfunction, as well as some course grained details of the containing mechanism to understand where the fault is located. Yet, most information about structural and behavioral specifics of malfunctioning components/sub mechanisms, and their containing mechanisms, is left out (Hawkins and Woollons 1998; Bell et al. 2007).[2]

Consider, again, by way of example, the Functional Interpretation Language (FIL) methodology for malfunction analysis and explanation (Bell et al. 2007). In FIL, functions are effect functions and represented in terms of their *triggers* and *effects*. Triggers describe input states that actuate physical behaviors which result in certain (expected) effects. For instance, the function description "de-press_brake_pedal"-"red_stop_lamps_lit" of a car's stop light (p. 400). This description is a summary of some salient features of (manipulating) such artifacts; depressing the brake pedal will, if the system functions properly, result in the lighting of the stop lamps.

According to Bell et al. (2007) such trigger and effect representations serve two explanatory ends in malfunction analyses: firstly, they *highlight* relevant behavioral features of a given artifact, i.e., effects, and, simultaneously, provide the means to *ignore* less relevant or irrelevant behavioral features, i.e., physical behaviors underlying these effects; secondly, they support assessing which components are malfunctioning (pp. 400–401).

For instance, the trigger-effect representation "depress_brake_pedal"-"red_-stop_lamps_lit" highlights the input condition of a pedal being depressed, and the resulting desired effect of lighted lamps, yet ignores the structural and behavioral specifics of the brake pedal and stop lamps, such as the pedal lever and electrical circuit mechanisms, as well as the energy conversions—e.g., mechanical energy

---

[2]That is, structural and behavioral characteristics are considered irrelevant in a first round functional analysis of malfunction. After this analysis, more detailed behavioral models of components and their behaviors are used for identifying specific explanatorily relevant structural and behavioral characteristics of malfunctioning components/sub mechanisms (Bell et al. 2007). However, immediately specifying these details in functional models is taken to result in listing a lot of irrelevant details.

conversions into electricity—that are needed to achieve this effect. Such representations only highlight those features that are considered explanatorily relevant to assess malfunctioning systems, and omit reference to physical behaviors/energy conversions by which desired effects are achieved.

Secondly, such trigger-effect descriptions support comparing normally functioning technical systems with malfunctioning ones (Bell et al. 2007). Trigger-effect descriptions support assessing whether the expected effects in fact obtain, and, if not, which and how components are malfunctioning (Bell et al. 2007). A normally functioning artifact, say the car's stop lights, has both a trigger and an effect occurring; the brake pedal is depressed and the stop lights are lit. Trigger-effect descriptions support analysis of two varieties of malfunction. First, a trigger may occur, yet fail to result in the intended effect. Say, the brake pedal is depressed, yet the stoplights are not on. Second, a trigger may not be occurring, yet the effect is nevertheless present. Say, the brake pedal is not depressed, yet the stoplights are on (see Bell et al. 2007). Such analysis of the actual states of triggers and effects allows one to focus on the most likely causes of failure (Bell et al. 2007). Say, if the pedal is depressed and the lights fail to ignite, first likely causes to investigate may be whether the electrical circuits in the lights are broken or the 'on/off' connection between the brake and electrical circuitry (connected to the lamp) is damaged. On the other hand, if the pedal is not depressed and the lights are lit, a first likely cause to investigate may be whether the 'on/off' connection between the brake and the electrical circuitry is damaged. To support more detailed malfunction analyses, functions are often decomposed into sub functions in FIL. An example of a functional decomposition of a two-ring cooking hob is given in Fig. 2.3.

The usage of effect functions and functional decompositions in FIL is the optimal choice given that function descriptions are used to black-box or suppress reference to unwanted behavioral and structural details. Effect function descriptions only highlight the relevant difference making properties with respect to malfunctioning artifacts, whereas more elaborate behavior function descriptions include irrelevant details such as, say, the thermal energy generated when lamps are lit.

Effect function descriptions also prove the optimal choice in the third explanation-seeking context that we consider: type level capacity explanation.

2.2.5 Abstraction, Generality, and Type Level Capacity Explanation

Explanatory models specified in terms of behavior function descriptions, which typically are represented by operations-on-flows (e.g. Hirtz et al 2002; Otto and Wood 1998, 2001; Pahl and Beitz 1988), as in the reverse engineering case, are fairly precise and complete when measured against models solely specified in terms of effect function descriptions, which typically are represented by verb-noun pairs (e.g., Bell et al. 2007; Deng 2002; Kitamura et al. 2005). The omission of details in

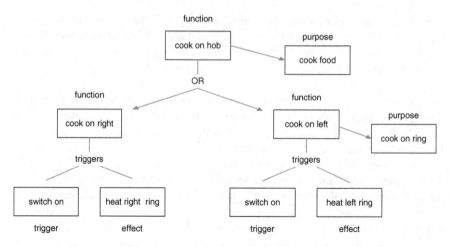

Fig. 2.3 Effect functional decomposition of a two-ring cooking hob (adapted from Bell et al. 2007)

explanatory models has important advantages, as discussions of abstraction and generality make clear (Weisberg 2007; Levy and Bechtel 2013); it makes such abstract models suitable for describing and explaining a larger class of technical systems, i.e., for type level capacity explanation rather than capacity explanation of individual tokens (as in the reverse engineering case). The Functional Concept Ontology (FCO) method for design and design knowledge management gives a good illustration of this point (Kitamura et al. 2005).

In a nutshell, the method uses knowledge bases in which, amongst others, functional descriptions of types of extant technical systems are archived, as well as part-whole relations between functions and sets of sub functions that compose 'upper level' functions. Functional descriptions in this method are descriptions of effect functions (van Eck 2011). The part-whole relations are 'enriched' with specifications of general technological principles by which sets of sub functions compose or achieve 'upper level' functions. These technological principles are called 'ways of achievement' (Kitamura et al. 2005). An example of an effect functional decomposition of a type of heavy duty stapler is given in Fig. 2.4.

By solely specifying effect functions and abstract, general technological principles, and omitting details about the precise manner in which materials, energies, and signals are processed, i.e., by not referring to behavior functions, such models are useful to capture the operation of types of mechanisms rather than individual tokens mechanisms. They focus on common features across token systems only, and omit reference to material energy and signal conversions that may differ across these token systems. They can be invoked to explain complex capacities of types of technical systems, here a type of heavy duty stapler, and such explanations are constructed using effect functions and functional decompositions.

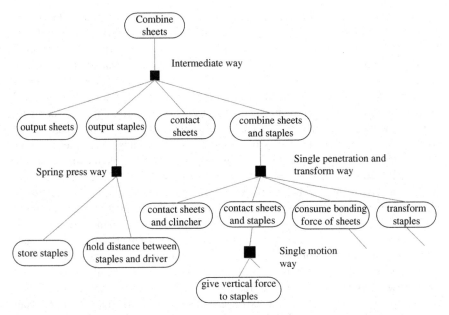

Fig. 2.4 Effect functional decomposition of a stapler. Functions are described in *ovals, black squares* refer to ways of achievement (adapted from Ookubo et al. 2007, p. 9, Fig. 3b.)

Both precision and generality are, as in other scientific domains, important in engineering: precise models offer in-depth understanding of the manner in which specific technical systems work and thus offer the means to adjust specific details in redesign phases in order to improve system functionality; more abstract and general models explain how types of technical systems operate. Such models are useful in (re) design contexts where predominantly knowledge on functional organization drives the initial design phase, and component-solutions are not considered in the initial phase of function specification, so as to consider different solution variants for these functional organizations (van Eck 2015b).

Since these desiderata of precision and generality are difficult to meet with single models, behavior functions and functional decompositions are used when precision is required and effect functions and functional decompositions are used when generality is needed. In engineering design, specific notions of function and functional decomposition are tailor-made depending upon the explanatory and/or design task at hand.

2.2.6 Capturing Mechanistic Explanation in Engineering Science: Pluralism About Mechanistic Role Functions

The upshot of these three cases is that explanations in engineering (as in every science of course) are constructed relative to explanatory objectives and, importantly, that the level of detail included in these explanatory models hinges on specific concepts of technical function. This latter feature marks a relevant distinction with the manner in which role function ascription and mechanism individuation is understood in the literature on mechanistic explanation in the life sciences. Engineering scientists simplify or increase the details of explanations—functional decompositions—depending on the explanatory purpose at hand, and these adjustments are made using specific concepts of technical function (compare e.g., Figs. 2.2 and 2.3, or Figs. 2.2 and 2.4). In the context of reverse engineering explanation of complex capacities of token technical systems, elaborate or 'complete' descriptions of mechanisms are provided, in terms of behavior functions and functional decompositions, to answer the question how a specific technical system exhibits a given overall behavior. In malfunction explanation, less elaborate 'sketches' of mechanisms are provided in terms of effect functions and functional decompositions, referring only to some mechanistic features, namely those difference making factors that mark the *contrast* between normal functioning and malfunctioning technical systems. Finally, when explaining complex capacities of types of technical systems, abstracting away from specific details of individual token cases, effect functions and functional decompositions are invoked. So, depending upon explanatory context, mechanisms are individuated in different ways using different conceptualizations of function in engineering science. Function ascription thus again proves highly relevant, both for type and token level capacity explanation and for malfunction explanation. Importantly, neither function conceptualization in itself accommodates both ways in which mechanisms are functionally individuated in engineering science. Behavior and effect function ascriptions are invoked to individuate mechanisms in different ways depending on the task at hand.

However, this distinction in functional individuation, and its reliance on different subtypes of function, is blurred in a perspective that understands mechanism individuation and mechanistic explanation in terms of mechanistic role function ascription simpliciter. The concept of mechanistic role function, an activity that makes a contribution to the workings of a mechanism of which it is a part, admits of two interpretations in the context of engineering science: behavior function on the one hand and effect function on the other. So the point is that in order to arrive at empirically informed understanding of explanatory practices in engineering, and at consistency of the general structure of mechanistic explanation with these practices, regimenting the concept of role function into domain-specific engineering concepts of behavior and effect function, i.e., sub types of role function, is needed.[3]

---

[3]Note that behavior and effect descriptions of function describe, in different ways, the contributions of components to mechanisms of which they are a part. The distinction between behavior and effect function thus is not to be conflated with the distinction between a mechanism description and a description of a mechanisms' overall activity. Neither is the behavior-effect function distinction to be conflated with the distinction between 'isolated' and 'contextual' descriptions of an entity's activity (Craver 2001): isolated descriptions describe activities without taking into account the mechanisms in which they are situated; contextual descriptions describe activities in terms of the

I now briefly consider another facet of the relationship between mechanistic explanation and engineering that has received little sustained analysis: the usage of engineering principles in the construction of mechanistic explanations in systems biology. Here we will see again that the distinction in subtypes of role function is relevant; the manner in which biological mechanisms are individuated in engineering terms, hinges on specific engineering conceptualizations of function. Specifically, effect function descriptions are used to describe and explain biological mechanisms in abstract, truncated fashion.

2.3   Explanation by Effect Functional Decomposition: Where Engineering and Systems Biology Meet

2.3.1   Engineering and Mechanistic Explanation in System Biology: The E. coli Heat Shock Case

Although philosophy, it seems, is only recently picking up on the fruitful cross-talk between engineering and systems biology (cf. Braillard 2015), engineers and systems biologists alike have been stressing the conceptual ties for more than a decade (Hartwell et al. 1999; Lazebnik 2002). With biological data about complex biological systems exploding during the last twenty years or so, due to (functional) genomics projects and the like, opportunities to understand complex biological systems in far greater detail became available. Yet cashing out that promise also signaled the need for new tools that enabled massive data analysis and integration in order to build explanatory models of these complex systems with a scale and complexity hitherto unknown. Here is where, amongst others, engineering tools came in. For instance, decomposition and control principles governing the construction of engineering systems are now being used to characterize complex biological systems (Tomlin and Axelrod 2005).

A case in point is research by El-Samad et al. on the mechanism(s) to counter heat shock in *E. coli* (El-Samad et al. 2005; cf. Tomlin and Axelrod 2005; Braillard 2015). Heat shock response is a widely conserved response of cells to cope with environmental stress brought about by unusual increases in temperature, involving the induced expression of heat shock proteins. Such temperature increases can damage proteins by breaking down their tertiary structures. Heat shock proteins come in two varieties and mitigate this effect in two different ways: molecular chaperones do so by refolding denatured proteins and proteases by degrading denatured proteins. If the response is sufficiently swift and massive, cell death can

---

(Footnote 3 continued)

mechanistic contexts in which they are situated and to which they contribute. Both behavior and effect functions are of the contextual variety, describing contributions of components to the mechanisms of which they are a part.

be prevented by protein repair and/or removal of damaged proteins. The response needs to be tightly controlled in the sense that it is only activated in case of heat shock, since the response is highly energy consuming and would make too high energy demands if heat shock proteins would be produced all the time. Cells thus must maintain a delicate balance between the protective effect of heat shock protein production and the metabolic cost of overproducing these proteins. In *E. coli*, the RNA polymerase cofactor φ^{32} promotes the transcription of heat shock proteins. After heat shock stress—temperature increase—φ^{32} activity increases, resulting in the transcription of specific heat shock gene promoters, which initiate transcription of genes, which in turn encode specific heat shock proteins—chaperones and proteases. This heat shock protein expression, when appropriate, prevents cell death. This mechanism uses both feed forward and feedback loops that process information about temperature and the folding state of proteins in the cell. φ^{32} activity is crucial in all this and depends on a feed forward mechanism that senses temperature and controls φ^{32} transcription, and feedback regulatory mechanisms that register the folding levels of proteins (levels of denatured cellular protein) and degrade φ^{32}. These regulatory feedback mechanism are crucial to ensure that φ^{32} synthesis, activity, and stability is brought back to normal levels after a sufficient number of heat shock proteins have been produced and the threat to cell death is averted.

El-Samad et al. (2005) constructed a quantitative, mathematical model of the heat shock response in order investigate the dynamical, mechanistic organization that sustains the heat shock response. They came up with an elaborate mathematical model consisting of 31 equations and 7 parameters. To make the model computationally tractable and be able to pose and answer questions about the dynamical, mechanistic organization of the system, the original model had to be trimmed down. As Braillard (2015) stressed, control engineering principles played an important heuristic role in this model reduction, i.e., abstraction, and thus in the discovery of the mechanism' core organizational features that sub serve its overall regulatory behavior. The close analogy between engineered systems and biological ones with respect to functional modular organizations sub serving regulatory processes made this possible. As El-Samad et al. (2005) explain:

> Control and dynamical systems theory is a discipline that uses modular decompositions extensively to make modeling and model reduction more tractable. Because biological networks are themselves complex regulation systems, it is reasonable to expect that seeking similarities with the functional modules traditionally identified in engineering schemes can be particularly useful. (El-Samad et al. 2005: 2737).

In control engineering, decomposition into functional modules (modules defined in terms of their effect-role functions) often begins with identification of the process to be regulated called the 'plant' (cf. Lind 1994), for instance altitude regulation of an airplane or temperature regulation of a thermostat. Modules of the system that contribute to the regulation are described in terms of their contributing functions, the most common of which are 'sensors', 'detectors', 'controllers', 'actuators', and 'feed forward' and 'feedback' signals. For instance, in a simple heating system, the plant is the temperature regulation process, which is achieved, inter alia, by a sensor module which measures ambient temperature, calculates the deviation from the desired temperature and feeds this information into the thermostat (controller). The thermostat then outputs signals that are send to an actuator (heat fuel valve) that generates an actuation signal (e.g., fuel to furnace) that corrects deviation from the desired temperature. The sensor module again measures the ambient temperature and, if needed, feeds back information on temperature deviations to the controller, and so on.

El-Samad et al. (2005) applied this control engineering perspective to the *E. coli* heat shock response system. In this application, the protein folding task (the refolding of denatured proteins) is taken to be the process to be regulated (plant), the feed forward signal (send by a sensor) is the temperature dependent translational efficiency of σ^{32} synthesis, the controller is the level of σ^{32} activity, chaperones function as actuators of the plant (the actuated plant input is the number of molecular chaperones), and sensors measure plant output (amount of denatured protein), which in turn is fed back to the controller.

This decomposition allowed El-Samad et al. (2005) to construct a simplified model consisting of just 6 equations and 11 parameters in which each equation describes aspects of the behavior of a module. They remark:

> This model provides useful insight into the heat shock system design architecture. It also suggests a mathematical and conceptual modular decomposition that defines the functional blocks or submodules of the heat shock system. This decomposition is drawn by analogy to manmade control systems and is found too constitute a canonical blueprint representation for the heat shock network. (El-Samad et al. 2005: 2736)

What we here thus see is that analogical reasoning with respect to regulation processes and the functional architecture sub serving these processes in engineered and biological systems, led to a functional modular decomposition of a biological system in terms of effect function descriptions that laid bare core organizational features of the system by which it produces regulatory behavior. Engineering tools —modular decompositions specified in terms of effect functions—here serve as a discovery heuristic for a mechanism' core organizational features that sub serve its overall regulatory behavior (cf. Braillard 2015) This usefulness of engineering concepts, i.e., modular decompositions in terms of effect functions, is not specific to the *E. coli* case, but generalizes to a variety of cases (cf. Tomlin and Axelrod 2005) and suggests a general discovery heuristic:

> If the heat shock mechanism can be described and understood in terms of engineering control principles, it will surely be informative to apply these principles to a broad array of cellular regulatory mechanisms and thereby reveal the control architecture under which they operate (Tomlin and Axelrod 2005: 4220).

Analysis of engineering function and explanation has more to offer. In concluding this chapter, I revisit the engineering explanation-seeking contexts from Sect. 2.2 and suggest that these illustrate the complementarity of two allegedly competing perspectives, 'completeness and specificity' (Craver 2007) and 'abstraction' (Levy and Bechtel 2013), on the explanatory power of mechanistic explanations. And that they pull in opposite directions in the context of malfunction explanation and, hence, that a novel desideratum is required to handle this explanatory context.

2.4 Explanatory Power: Rethinking the Explanatory Desiderata of 'Abstraction' and 'Completeness and Specificity'[4]

According to one influential perspective, the power of mechanistic models is (almost) always increased when these refer to both functional and structural features of mechanisms (Machamer et al. 2000; Craver 2007). On the counterview, mechanistic models have in certain contexts more explanatory traction when reference to structural aspects of mechanisms is suppressed. Models that solely describe functional characteristics, i.e., causal relations between components, explain better how organization impacts the behavior of mechanisms (Levy and Bechtel 2013). The engineering cases presented here allow for a more fine-grained understanding of the relationship between these views: rather than being in competition, they emphasize different explanatory virtues that hold in different explanation-seeking contexts.

I have argued elsewhere that differences between these two (allegedly) competing perspectives on the explanatory power of mechanistic explanations, 'completeness and specificity' (Craver 2007) and 'abstraction' (Levy and Bechtel 2013), essentially boil down to differences in the notions of difference making endorsed in these accounts and that they are in fact not in competition (van Eck 2015a). They are rather suitable for different explanation-seeking contexts. Whereas abstraction dictates that mechanistic explanations should only list the 'primary factors' responsible for the occurrence of system function, 'completeness and specificity' prescribes that in addition to primary ones also 'higher order factors' should be described, which concerns factors influencing the precise manner in which a system function occurs or those sub serving the primary factors. The engineering cases gives an empirical illustration of this 'complementarity view'.

2.4.1 Malfunction Explanation: Local Specificity and Global Abstraction

In the context of reverse engineering explanation presented here, i.e., token level capacity explanation, engineers take details to matter: elaborate behavior functional decompositions, and related component models, are constructed to describe the mechanisms of specific artifacts, via the breaking down of artifacts

---

[4]This section draws on van Eck (2015a).

component-by-component and assessing the effects of single component removals on their overall behaviors. This perspective agrees with the 'completeness and specificity' view on mechanistic explanations. In the model of the reverse engineered electrical screwdriver in Fig. 2.2, for instance, both factors that make a difference to the occurrence of the screwdriver's overall behavior are listed, such as 'supply electricity' and 'convert electricity to torque', as well as factors that affect the way in which this behavior is manifested, such as 'dissipate torque' into 'heat' and 'noise' flows, and 'allow rotational degrees of freedom' (the latter concerns controlling the movement of materials along a specific degree of freedom (Stone and Wood 2000), here appropriate hand positions for correct functioning of the screwdriver).

Such primary and higher order details matter given that the reverse engineering explanation ultimately is in the service of redesign purposes: identifying components that function sub-optimally in a reverse engineered artifact and subsequent optimization in redesigned artifacts. The manner in which a particular technical system exhibits a given piece of behavior then becomes important. For instance, in an empirical example of the reverse engineering of an electric wok and its subsequent redesign, structural features of components affected the precise manner in which temperature distribution across the wok's bowl was manifested, and modifications of these features were needed to optimize temperature distribution across the bowl; the electric heating elements of the wok, such as a bimetallic temperature controller, were housed in too narrow a circular channel and optimized in the redesign phase (Otto and Wood 1998).

The abstraction perspective, on the other hand, is suitable in the context of type level capacity explanation. There the omission of details concerning the precise manner in which materials, energies, and signals are processed sub serves the description and explanation of the workings of (multiple) types of technical systems, rather than specific token systems. Such models only require the specification of primary factors that affect the occurrence of specific complex capacities. For instance, the capacity of heavy duty staplers to 'connect sheets' (cf. Fig. 2.4). Higher order details are not needed, since these are or might be specific to particular tokens systems.

At first glance, it seems that the abstraction perspective is also better suited to capture malfunction explanation. In that context, as we saw, engineers advance the maxim that 'less is more' when it comes to adequate explanations. Closer inspection however reveals that in this explanatory context 'abstraction' and 'completeness and specificity' pull in opposite directions (van Eck 2015a).

To see this, consider that in order to understand how a malfunctioning component or sub mechanism makes a difference to the *occurrence* of a specific system level malfunction, one needs to know how the failing component or sub mechanism is situated within a mechanism that underlies normal functioning. That is, malfunctions are identified against a backdrop of normal mechanism functioning (cf. Thagard 2003; Moghaddam-Taaheri 2011). This is required to explain the contrast drawn in the explanandum—why malfunction, rather than normal function.

This also happens in FIL, in which function descriptions and functional decomposition models in terms of trigger-effect descriptions are used to specify normal functioning, and to provide the context against which to assess specific malfunctions, such as a trigger that occurs yet fails to result in an expected effect—say, a cooking hobs' switch that is on but does not result in the heating of a ring (Bell et al. 2007). Such contrastive factors that explain the contrast drawn in the explanandum, i.e., make the difference, between malfunction and normal function are primary ones that underlie the occurrence of the specific system-level malfunction in question. Say, in the above example, the electrical circuitry connected to the ring that is damaged as a result of which the ring does not heat, and food cannot be heated. Also the details on normal functioning that are needed to understand why the factor(s) cited in the explanans, e.g., a broken electrical wiring, is a contrastive one, concerns primary factors that underlie normal functioning. Since fact and foil in the contrastive explanandum concern the occurrence of malfunction and function, respectively, the factors needed to understand which part(s) of the mechanism malfunction and which ones function normally should be primary ones as well. Information on the precise manner in which mechanisms normally manifest their functions is irrelevant here. Knowing that rings of cooking hobs normally heat when switches are thrown is sufficient to understand that when this trigger-effect relation does not obtain, a malfunction occurs.

Also, it suffices to describe properly functioning parts of mechanisms in abstract fashion, i.e., in terms of functionally characterized components and their functions, since their job is only to highlight where in the mechanism a malfunctioning component or sub mechanisms is located. Listing structural features, such as size and shape, is irrelevant here for what matters is knowing what these components/sub mechanisms (normally) do. I here label the constraint to specify common features of functioning and malfunctioning mechanisms in terms of functionally characterized components and their functions, '*global abstraction*'. However, the contrastive factor(s) that makes the difference to the occurrence of a specific system-level malfunction often will have to be described in more elaborate fashion and its description will, in addition to functional characteristics, also refer to structural features. The manner in which a component is, say, broken or worn often does make a difference to the occurrence of a system level malfunction. A rupture in the electrical wiring of the cooking hob, for instance, which leads to failure of the ring to heat. Here specificity with respect to structural features is needed as well. I label this constraint to describe both functional and structural characteristics of contrastive difference makers, '*local specificity*' (both to set it apart from 'global abstraction', and from 'completeness' in the sense of specifying both primary and higher order factors; 'local specificity' as I understand it here concerns primary factors only).[5]

---

[5]This is in keeping with engineering practice. After a first round functional analysis of malfunction, more detailed behavioral models of components and their behaviors are used in FIL for assessing specific structural characteristics of malfunctioning components (Bell et al. 2007).

Malfunction explanations thus require a format in between 'completeness and specificity' and 'abstraction': they require *local specificity* with respect to descriptions of malfunctioning components/sub mechanisms and *global abstraction* with respect to descriptions of the mechanisms in which the component/sub mechanism failures are placed. This analysis extends current thinking about the explanatory power of mechanistic explanations by spelling out a novel desideratum for malfunction explanations. The lesson is that in this context, explanations that contain local specificity and global abstraction are better than either complete or abstract mechanistic explanations. And, as we saw, in the context of engineering science, depending on the richness that is required of explanations, specific concepts of technical function and functional decomposition are invoked. The examples of reverse engineering explanation/token level capacity explanation analyzed here use behavior functions and functional decompositions, whereas malfunction explanations and type level capacity explanations are procured in terms of effect functions and functional decompositions.

A further question emerges: is 'local specificity and global abstraction' a desideratum only for malfunction explanations of technical systems, or does it also apply to malfunction explanations in other scientific domains, like biology? I argue below that explanations of biological malfunctions also best exhibit 'local specificity and global abstraction'.

2.4.2 Malfunction Explanation in Biology

Also in the case of explaining biological malfunction, I take it that explanations that are locally specific and globally abstract are the optimal ones. Consider, for instance, impaired blood circulation in the circulatory system.[6] Malfunction explanations, of course, should single out those steps—entities engaging in activities—in the circulatory system's mechanism(s) that cause the circulation of blood to be impaired, i.e., make a difference to whether or not impaired blood circulation occurs. In the case of impaired blood distribution, the cause may be that blood transport is disrupted in particular vessels as a result of thrombosis in those vessels. The description of these contrastive factors—damaged vessels due to thrombosis— often will have to be described in elaborate fashion, i.e., in terms of both functional and structural specifics. In our example, it is relevant to know that the damaged vessels fail to perform their function of transporting blood. Yet the manner in which those vessels are damaged, and thus fail to perform their function(s), also makes a difference to the occurrence of impaired blood circulation. When the vessels are only slightly damaged they may still perform their function of transporting blood, so it is relevant to know the nature of the damage, i.e., the manner in which

---

[6]I adapt this example from Nervi (2010).

structural features of the vessels are deformed. Here, deformations due to thrombosis. Local specificity thus applies to descriptions of such contrastive difference makers.

And, again, to explain the contrast drawn in the explanandum—why malfunction, rather than normal function—one also needs to know how the failing component or sub mechanism is situated within a mechanism that underlies normal functioning, since malfunctions are identified against a backdrop of normal mechanism functioning (cf. Thagard 2003; Moghaddam-Taaheri 2011). However, descriptions of the relevant properly functioning parts of mechanisms can be given in abstract terms—functionally characterized components and their functions—since their job is only to highlight where in the mechanism a malfunctioning component or sub mechanisms is located. It suffices to know that, say, the cardiac muscle engages in coordinated contraction, that blood is ejected from the ventricles into the aorta and the arterial system, etc. Further detailing of structural specifics, say, the precise shape or size of the cardiac muscle has no added value for locating the fault(s) in the mechanism. So, the desideratum of 'local specificity and global abstraction' is not restricted to malfunction explanations of technical systems, but applies more broadly to malfunction explanations in the biological domain as well.

References

Bechtel, W., & Abrahamson, A. (2005). Explanation: A mechanist alternative. *Studies in History and Philosophy of Biological and Biomedical Sciences, 36*, 421–441.

Bell, J., Snooke, N., & Price, C. (2007). A Language for functional interpretation of model based simulation. *Advanced Engineering Informatics, 21*, 398–409.

Braillard, P.A. (2015). Prospects and limits of explaining biological systems in engineering terms. In P.A. Braillard & C. Malaterre (Eds.), *Explanation in biology* (pp. 319–344). Springer.

Chandrasekaran, B., & Josephson, J. R. (2000). Function in device representation. *Engineering with Computers, 16*, 162–177.

Craver, C. F. (2001). Role functions, Mechanisms, and Hierarchy. *Philosophy of Science, 68*, 53–74.

Craver, C. F. (2007). *Explaining the brain: Mechanisms and the mosaic unity of neuroscience.* New York: Oxford University Press.

Csete, M. E., & Doyle, J. C. (2002). Reverse engineering of biological complexity. *Science, 295*, 1664–1669.

Darden, L. (2006). *Reasoning in biological discoveries.* Cambridge: Cambridge University Press.

Deng, Y. M. (2002). Function and behavior representation in conceptual mechanical design. *Artificial Intelligence for Engineering Design, Analysis and Manufacturing, 16,* 343–362.

El-Samad, H., Kurata, H., Doyle, J. C., Gross, C. A., & Khammash, M. (2005). Surviving heat shock: control strategies for robustness and performance. *PNAS, 102*(8), 736–2741.

Erden, M. S., Komoto, H., van Beek, T. J., D'Amelio, V., Echavarria, E., & Tomiyama, T. (2008). A review of function modeling: approaches and applications. *Artificial Intelligence for Engineering Design, Analysis and Manufacturing, 22,* 147–169.

Glennan, S. (2005). Modeling mechanisms. *Studies in the History and Philosophy of the Biological and Biomedical Sciences, 36*(2), 375–388.

Goel, A. K. (2013). A 30-year case study and 15 principles: Implications of an artificial intelligence methodology for functional modeling. *AIEDAM, 27*(3), 203–215.

Hawkins, P. G., & Woollons, D. J. (1998). Failure modes and effects analysis of complex engineering systems using functional models. *Artificial Intelligence in Engineering, 12*(4), 375–397.

Hartwell, L. H., Hopfield, J. J., Leibner, S., & Murray, A. W. (1999). From molecular to modular cell biology. *Nature, 402,* C47–C52.

Hirtz, J., Stone, R. B., McAdams, D. A., Szykman, S., & Wood, K. L. (2002). A functional basis for engineering design: reconciling and evolving previous efforts. *Research in Engineering Design, 13,* 65–82.

Illari, P., & Williamson, J. (2010). Function and organization: Comparing the mechanisms of protein synthesis and natural selection. *Studies in History and Philosophy of Biological and Biomedical Sciences, 41,* 279–291.

Illari, P., & Williamson, J. (2012). What is a mechanism? Thinking about mechanisms across the sciences. *European Journal for Philosophy of Science, 2,* 119–135.

Kitamura, Y., Koji, Y., & Mizoguchi, R. (2005). An ontological model of device function: Industrial deployment and lessons learned. *Applied Ontology, 1,* 237–262.

Lazebnik, Y. (2002). Can a biologist fix a radio?—Or, What I learned while studying apoptosis. *Cancer Cell, 2,* 179–182.

Levy, A., & Bechtel, W. (2013). Abstraction and the organization of mechanisms. *Philosophy of science, 80,* 241–261.

Levy, A. (2014). Machine-likeness and explanation by decomposition. *Philosopher's imprint, 6,* 1–15.

Lind, M. (1994). Modeling goals and functions of complex industrial plants. *Applied Artificial Intelligence, 8,* 259–283.

Machamer, P. K., Darden, L., & Craver, C. F. (2000). Thinking about mechanisms. *Philosophy of Science, 57,* 1–25.

Moghaddam-Taaheri, S. (2011). Understanding Pathology in the context of physiological mechanisms: The practicality of a broken-normal view. *Biology and Philosophy, 26,* 603–611.

Nervi, M. (2010). Mechanism, malfunctions and explanation in medicine. *Biology and Philosophy, 25,* 215–228.

Ookubo, M., Koji, Y., Sasajima, M., Kitamura, Y., Mizoguchi, R. (2007). Towards interoperability between functional taxonomies using an ontology-based mapping. In *Proceedings of the International Conference on Engineering Design (ICED 07),* 28-31 Aug 2007, Paris, France: 1–12.

Otto, K. N., & Wood, K. L. (1998). Product evolution: A reverse engineering and redesign methodology. *Research in Engineering Design, 10,* 226–243.

Otto, K. N., & Wood, K. L. (2001). *Product design: Techniques in reverse engineering and new product development.* Upper Saddle River NJ: Prentice Hall.

Pahl, G., & Beitz, W. (1988). *Engineering design: A systematic approach.* Berlin: Springer.

Stone, R. B., & Wood, K. L. (2000). Development of a Functional basis for design. *Journal of Mechanical Design, 122,* 359–370.

Stone, R. B., Wood, K. L., & Crawford, R. H. (1998). A heuristic method to identify modules from a functional description of a product. *ASME proceedings,* 1–21.

Stone, R. B., Wood, K. L., & Crawford, R. H. (2000). A heuristic method for identifying modules for product architectures. *Design Studies, 21,* 5–31.

Thagard, P. (2003). Pathways to biomedical discovery. *Philosophy of Science, 70,* 235–254.

Tomlin, C. J., & Axelrod, J. D. (2005). Understanding biology by reverse engineering the control. *PNAS, 102*(12), 4219–4220.

van Eck, D. (2010). On the conversion of functional models: Bridging differences between functional taxonomies in the modeling of user actions. *Research in Engineering Design, 21*(2), 99–111.

van Eck, D. (2011). Supporting design knowledge exchange by converting models of functional decomposition. *Journal of Engineering Design, 22*(11–12), 839–858.

van Eck, D. (2015a). Mechanistic explanation in engineering science. *European Journal for Philosophy of Science, 5*(3), 349–375.

van Eck, D. (2015b). Validating function-based design methods: An explanationist perspective. *Philosophy and Technology, 28*, 511–531.

Vermaas, P. E. (2009). The Flexible Meaning of Function in Engineering, *Proceedings of the 17th International Conference on Engineering Design (ICED 09)*:vol. 2. 113–124.

Weisberg, M. (2007). Three kinds of idealization. *The journal of Philosophy, 104*(12), 639–659.

Chapter 3
Assessing the Roles of Design Representations: Counterfactual Understanding and Technical Advantage Predictions

Abstract In this chapter I elaborate two important roles of design representations in terms of concepts and insights from the philosophical literature on explanation: design representations as means for counterfactual understanding, and for articulating predictions concerning technical advantages. Examples from the functional modeling literature are used to illustrate these roles.

Keywords Design representation · Counterfactual understanding · Technical advantage

In the previous chapters we were concerned with representations and explanations of extant artifacts. Here we will be concerned with the (predictive and other) roles of design representations of to-be-built artifacts, and we will comment on the goodness of such design representations. Chapter four will be devoted to testing or assessing the goodness of representations of extant artifacts in more detail.

Understanding the role(s) of design representations is one of the main issues in philosophy of design (Galle 1999). Per Galle analyzed the topic in detail and discussed such roles against the backdrop of the (notorious) 'problem of the absent artifact'. He argued that a viable account of designing should address the 'problem of the absent artifact'. The problem concerns the nature of design representations. Specifically, the question how one can utter true statements, in terms of design representations, about artifacts when these artifacts are, in the design phase, still non-existent. Galle attempted to tackle the problem, and developed an account of the roles of design representations in terms of a 'solution' to the problem; in this account, design representations serve as a basis to explore and communicate 'truths' about designs.

In this chapter I expose the (alleged) 'problem of the absent artifact' as a pseudo-problem, and in effect elaborate different and more plausible roles of design representations than Galle envisages. I argue that design representations are not means for the production of truth-apt assertions. This dissolves the 'absent artifact problem'. I elaborate an alternative view, in terms of concepts and insights from the philosophical literature on explanation, according to which design representations are means for counterfactual understanding and for articulating predictions concerning *technical advantages*. Examples from the functional modeling literature (some of which we already encountered in the previous two chapters) are used to illustrate these roles.

3.1 Introduction

Part of any philosophical domain of inquiry concerns analysis of the nature of key concepts in its domain. For instance, in the philosophy of science literature there is an ongoing attempt to understand the nature of scientific explanation, e.g., whether explanation is ontic or epistemic (cf. Craver 2007), and which formats best capture the structure of such explanations (e.g., covering law explanation, explanation by unification, or along causal-mechanical lines). Likewise, in the philosophy of design, the related key notions of 'designing' and 'design representation' have been the subject of detailed investigations (as well as the question whether design is a branch of science or not; I comment on this question in Chap. 4). Key tasks in understanding the nature of designing concern understanding what design representations are and which roles they fulfill in the design process (Galle 1999; Herbert 1993). Galle (1999, p. 58, 62) defended the view that designing is "the production of a design representation" and identified two essential roles of design representations: means for 'communication' and for 'exploration'. By his lights, in order to understand the nature and role of design representations, any account of designing should address what he calls the 'problem of the absent artifact':

> How can we (apparently) utter and communicate truths about things which are not there to make our propositions true? These questions, when asked of design representations, state what I shall call *the problem of the absent artefact* (Galle 1999, p. 66, italics in original; cf. Herbert 1993)

For instance, how to make sense of cases like this:

> the architect may truthfully tell his client that 'the house' he is designing complies with the fire safety regulations, even though there is not yet any house at hand to comply with anything (Galle 1999, p. 66)

More generally, how can design representations enable us, in the design phase, to produce true assertions about artifacts, *when these artifacts have not yet been produced/built?*

Galle attempted to address the problem by arguing that design representations do not refer to to-be-built artifacts, but are related to cognitive entities/ideas and that, rather than spatial-temporally located items in the world, ideas are the truth-makers of statements about designs. I argue that this move is deeply problematic for this notion of ideas as truth-makers cannot be justified and that, hence, in his account it is not possible to give a plausible interpretation of the idea that design representations serve roles in the exploration and communication of truth-apt statements about designs (cf. van Eck 2015).

In this chapter I furthermore argue that the 'problem' of the absent artifact is a wrong turn in the philosophy of design: it is not a pressing problem that is in need of a solution. I argue, pace Galle, that design representations are *not* means for 'exploring' and 'communicating' truths about designs. Design representations, and utterances based on them, can be subjected to evaluation in terms of a variety of norms such as 'generality', 'precision', and 'completeness' (van Eck 2015), yet 'truth' is not among them. This dissolves the 'absent artifact problem'. By getting the alleged problem out of the way, we can make headway on developing a more fruitful approach towards understanding the roles played by design representations in designing. I here aim to get grip on the roles of design representations in terms of concepts from the philosophical literature on scientific understanding, in particular the ideas of *counterfactual understanding* and *counterfactual comparison* (cf. Woodward 2003).

I argue in this chapter that design models or representations are better understood as 'vehicles' to procure (counterfactual) understanding of to-be-built artifacts in terms of offering answers to *what-if-things-had-been-different questions*, and as predictive devices to counterfactually compare the functional performance of extant artifacts with to-be-redesigned systems. Empirical examples from the engineering functional modeling literature are used to illustrate these roles of design representations.

The chapter is structured as follows. The absent artifact problem is introduced in the next section. The problem is dissolved in Sect. 3.3. The roles of design representations with respect to counterfactual understanding and prediction are discussed in Sect. 3.4. Section 3.5 concludes the chapter.

3.2 Design Representations and the Problem of the Absent Artifact[1]

Various proposals for understanding design thinking, like 'problem solving', 'information processing', 'decision making', and 'pattern recognition', have been proposed in the theoretical/philosophical literature on design. Endorsing Cross' (1992) assessment that such proposals, failed to capture all intricacies of design thinking, Galle develops a different approach.

Galle situates designing within the broader process of artifact production, which is construed, essentially, as a temporally ordered process of actions involving clients, designers, makers, and users. Such actions include, inter alia, clients producing design briefs, designers interpreting such briefs, designers producing design representations, and clients interpreting such representations. His proposal is to

---

[1]This section and Sect. 3.3 draw on van Eck (2015).

view designing as "the production of a design representation" (Galle, pp. 58, 62), which corresponds to one of the actions carried out by a designer in the artifact production process.[2] Since the production and interpretation of design representations are important actions comprising the artifact production process it is, of course, evident that:

> To understand designing we should understand the roles played by the design representation (Galle 1999, p. 62).

Galle stressed two essential roles of design representations: they enable "communication" and "exploration" (Galle 1999, p. 63). Design representations allow self-communication with the designer, i.e., the study and often subsequent revision of the design on the part of the designer, as well as communication with clients, makers, and users. These communicative acts correspond to some of the actions carried out during the artifact production process. Secondly, they allow exploration in the sense of being a means for answering hypothetical questions about to-be-built artifacts posed by designers, clients, and makers. Such questions may concern, inter alia, esthetic features, price/cost specifics, time and labor estimates, and structural specifics like load capacity of items (Galle 1999). Indeed, communication and exploration are part and parcel of most (engineering) design methods (e.g., Pahl and Beitz 1988; Otto and Wood 2001).

The idea that design representations sub serve communication and exploration is further fleshed out by Galle against the backdrop of the "problem of the absent artefact" (Galle, pp. 58, 62). For him, this (alleged) problem requires a "satisfactory solution" in order to explain what design representations are (cf. Herbert 1993). This problem concerns the questions 'to what do design representations refer?'; 'how is it possible to invoke design representations to assert truths (or falsehoods) about artifacts when these are, in the design phase, still non-existent?'; 'How can one communicate and explore 'truths' about designs?'.

The view that design representations must in some sense be descriptions of artifacts is immediately rejected by Galle since it then needs explaining how design representations can refer to non-existent things. Galle develops a different approach. Design representations do not refer to to-be-built artifacts, but are related to cognitive entities or ideas. Rather than spatial-temporally located items in the world, ideas about artifacts, "artifact-ideas" for short, are the truth-makers of statements about designs. Artifact production thus becomes a temporally ordered process of actions related to the production and interpretation of artifact-ideas, and designing now gets characterized as the production of a design representation, understood as "a thing (material entity) related to ideas by actions of interpretation and production" (Galle 1999, p. 74).

---

[2]Vermaas (2009) and Houkes and Vermaas (2010) also stress the important role of actions in designing (cf. Chap. 1).

More specifically, on this approach (Galle 1999, p. 75):

a design representation is a thing which the designer produces, driven by the designer's artifact-idea, while

(d) intending the client to interpret it (i.e. the thing) as being in accordance with the client's artifact-idea;

(e) intending the maker to interpret it and, driven by the idea thereby acquired or accessed (the maker's artifact-idea), to produce another thing (an artifact); and

(f) intending the designer himself to interpret the artifact as being in accordance with the designer's artifact-idea.

So the relevant relata are design representations qua things and artifact ideas of designers, clients, and makers, and clause (f) refers to the situation when an artifact has been made and it is assessed how the designer's idea of the to-be-made artifact 'matches' the artifact that has been made. The roles of design representations in 'exploration' and 'communication' are thus to be understood in terms of exploring and communicating truths about designs, which are ultimately justified in terms of artifact-ideas.

If we apply this account to the hypothetical architect-scenario described in the introduction, the remark that "the architect may truthfully tell his client that 'the house' he is designing complies with the fire safety regulations, even though there is not yet any house at hand to comply with anything" (Galle 1999, p. 66) can then be understood as a claim on the part of the architect that features of the design representation, e.g., complying with safety regulations, correspond with his/her artifact-ideas. His/her artifact-ideas make such assertions about the design come out true or false.

In the next section I argue that this approach does not solve the problem of the absent artifact: the proposed 'solution' leads to a vicious 'justification' problem since ideas cannot function as truth-makers of propositions in the sense that Galle intends. I furthermore argue that there is a quite simple solution for it: design representations and models are *not* means for exploring and communicating truths about designs. Hence the problem of the absent artifact is a pseudo-problem. This conceptual groundwork allows me to illustrate in section three some, what I take to be, very important roles of design representations, i.e., as means for counterfactual understanding and as predictive devices to articulate the expected technical advantages of redesigned systems compared with extant ones (in section three it will become clear how I understand 'technical advantage').

3.3 Exposing the Problem of the Absent Artifact as a Pseudo-Problem

Galle (1999) gives a succinct statement as to why he rejects the view that design representations must in some sense be descriptions of artifacts and opts for his alternative 'artifact-idea as truth maker' approach:

The other approach is to distrust linguistic intuition and maintain instead that what design representations 'really' describe or express (and what accounts for the truth of propositions expressed or implied by them) is not an as-yet-non-existent artifact of the material world, but rather an entity belonging to a mental or cognitive realm and existing there when we describe or express it; let us call it an *idea*, for short. If we could explain how an artifact is eventually produced in accordance with such an idea that foreshadows it, we should have escaped the problem of the absent artifact. [...] This second approach seems to imply, somewhat puzzlingly, that what 'really' complies with the regulations is *an idea* of a building, rather than a building! More generally, what makes a proposition true or false would be an idea rather than a thing; a cognitive rather than a non-cognitive entity. (Galle 1999, pp. 66–67).[3]

Prima facie, ideas as truth-makers of propositions based on design representations is a puzzling feature indeed. Nevertheless, Galle accepts it and defends this second approach to solving the problem of the absent artifact.[4] The question thus before us is: are ideas of designers apt to fulfill their truth-ascribing role?

I argue that t/his approach is problematic for two main (and related) reasons. First, ideas *cannot* function as truth-makers of propositions in the sense required by Galle to have a useful notion of a design representation, i.e., one that facilitates communication and exploration of alethic statements about designs between designers, clients, and makers. Secondly, more generally, it is nonsensical to ask after the truth conditions of propositions based on design representations in the first place, when the artifact in question has not yet been built. I consider these problems in turn.

Remember that design representations are defined by Galle, inter alia, as things produced by designers, which they intend to be in accordance with clients' artifact ideas, as well as interpreted such by makers that the artifacts they (eventually) produce are in accordance with designers' artifact ideas. As Galle puts it:

When producing his representation, the designer is driven not only by his artefact-idea (his idea of 'the artefact itself' as our language might mislead us to say), but also by an idea of how the client will interpret the design representation (see clause d of the definition), by an idea of how the maker will interpret it and react to it (clause e), and by an idea of how the designer himself will interpret the outcome of the maker's reaction (clause f) (1999, p. 77).

---

[3]Of course, talk such as 'the truth of propositions expressed by design representations' is plainly misguided. Alethic norms then would apply to representations themselves, rather than propositions expressed by agents in terms of them.

[4]Galle takes this to be not so puzzling after all, since we perceive the world through a veil of ideas, i.e., perception is indirect (1999, p. 81). Hence, ideas play a role in both the assignment of truth values to propositions about worldly states of affairs and propositions about non-existent items, such as non-existent artifacts. Of course, perception is idea—or theory-laden, yet the truth-makers in the former case are items in the world perceived in specific fashion, and in the latter ideas as such *without a referential link to items in the world*. These are very different things (also when one takes ideas as items in the world the difference holds, since in the one case the truth makers are *other* items in the world to which the ideas refer, which does not apply in the case of solely ideas).

However, there seems no way to guarantee such accordance between ideas of designers, makers, and clients. Consider some of Galle's own misgivings about 'matching ideas' between different agents in the design process:

> it would be incautious to assume that an idea can somehow be externalized by one agent and subsequently internalized by another agent as *the same* idea (or a 'true copy' of the original idea), for we have no way in which to ascertain sameness or similarity of ideas residing in, or accessed by, different minds (Galle 1999, p. 72, italics in original).

In related fashion, paraphrasing an objection of a colleague, Galle remarks:

> since ideas of other people are not accessible, and the definition [of a design representation] is stated in terms of ideas, it may not be possible to decide whether it applies to a given thing (1999, p. 76).

These misgivings tell against the idea that ideas about designs (always) match or accord between different agents in the design process. Nevertheless, he concludes:

> I should think, however, that it is possible for the designer, at least, to decide whether or not a thing he produces is a design representation according to the definition. More objective or 'public' criteria would be desirable, of course, but are hard to come by. (Galle 1999, p. 76)

The problem is that such objective or public criteria are not only desirable, but according to Galle's own definition of a design representation also essential to a useful notion of a design representation, i.e., one that facilitates communication and exploration of designs between designers, clients, and makers, since design representations are defined, inter alia, as things produced by designers, which they intend to be in accordance with clients' artifact ideas, as well as interpreted such by makers that the artifacts they (eventually) produce are in accordance with designers' artifact ideas. Yet such accordance between ideas of designers, makers, and clients cannot be guaranteed since "we have no way in which to ascertain sameness or similarity of ideas residing in, or accessed by, different minds". Hence, the 'private' nature of ideas limits the usefulness of Galle's definition of a design representation with respect to communication and exploration.

Far more important, with an eye to the absent artifact problem, such 'public' criteria are required to spell out in meaningful fashion how ideas can be invoked to assign truth values to propositions that are based on design representations. However, assuming the private nature of ideas, it becomes impossible to *inter-subjectively* establish the truth or falsity of propositions expressed in terms of design representations in unambiguous fashion, since true-false statements are (completely) relativized to particular agents. For instance, the architect in our example may assert that the house she is designing complies with the safety regulations since her ideas make the proposition true, whereas the client may take this assertion to be false given her different ideas. This (possibly) undermines meaningful design discourse between different agents. In sum, ideas cannot function as truth makers of propositions in the sense required by Galle to have a useful notion of a design representation. His approach thus fails to solve the problem of the absent artifact.

There is a much deeper problem however. The 'problem' seems insoluble or intractable when one insists on focusing on truth-conditions with respect to assertions based on design representations and designing, unless ... one recognizes that it is nonsensical to ask after the truth conditions of propositions based on design representations in the first place, when the artifact in question has not yet been built. By my lights, the question admits of no sensible answer; as long as an artifact has not been built, there seems no intelligible way to invoke a design representation to make a truth-apt/valued statement. 'Truth' is simply not something that one should ask of propositions based on design representations. Consider that the exploration and communication of 'truths' about designs in particular concerns *predictions* about designs (Galle 1999, 2008). However, predictions concerning future states of affairs have an indeterminate truth value (cf. Tweedale 2004). Certain states of affairs need come to pass before predictions turn out true or false. In my view, this is an essential feature for something to be a prediction in the first place. In our case at hand, the artifact needs to have been build before one can assign truth values to predictive assertions about its design (one may perhaps say that such assertions had a truth value all along, but these can only be assigned in retrospect, after the production of an artifact). So, the concern with truth conditions of propositions based on design representations when there is still no artifact to refer to is deeply misguided.

It is important here to distinguish between 'truth' or 'correctness' on the one hand and 'plausibility' on the other. It is entirely sensible of course to inquire into the *plausibility* of design predictions, i.e., the likelihood that they will prove correct, but this concerns assessment of design representations and assertions based on them in terms of *non-alethic* norms. For instance, is the design representation in agreement with certain relevant technological and/or scientific principles or regularities, such that assertions based on them do not violate key design principles? Say, are the structural specifics of the to-be-built house specified such that it is likely that the structure, when built, is able to carry certain loads and does not collapse on the spot? Such plausibility assessments are based on, amongst others, known technological/scientific principles and knowledge of the workings of extant artifacts (many more principles can be and are invoked of course; in our house design-case, legal norms pertaining to fire prevention, for instance).

Such plausibility assessments are part and parcel of designing and come in many varieties. In the context of functional modeling in engineering design, for instance, there are ongoing efforts to arrive at definitions of technologically sound and feasible definitions of technical functions that can be used in the construction of design representations of to-be-built technical artifacts (Stone and Wood 2000; Hirtz et al. 2002). Similarly, in redesign contexts, the functional performance of extant artifacts is (counterfactually) compared with possible alternative designs with the aim to come up with plausible improvements of these extant artifacts (Otto and Wood 1998, 2001; cf. Sect. 3). Yet, attempting to make truth-apt assertions in such design contexts is ill-advised since the truth-makers, build artifacts, are in this phase non-existent.

Galle (1999) and Herbert (1993) are on to an important issue here, i.e., that the absence of artifacts in the design phase urges us to analyze the status and role of design representations. However, accounts that attempt to develop such analyses in terms of truth-conditions are betting on the wrong horse. Design representations are not means for the production of truth-valued statements or the exploration and communication of truths about designs.

In the next section I present an alternative take on the issue and consider two important roles of design representations, to wit: as means for counterfactual understanding and as means to make predictions with respect to the functional performance of redesigned systems vis-à-vis original, extant ones. Functional modeling methods for designing serve as illustrating cases.

3.4 Elaborating Roles of Design Representations

3.4.1 Counterfactual Understanding

Stone and Wood's (2000) 'Functional Basis' (FB) method for designing gives a good illustration of the role of design representations in *counterfactual understanding*. In this method, designing starts by specifying an overall product function of an artifact to be designed, which is derived from customer needs. Such product functions are represented as operations on flows of materials, energies, and signals. The product function is then decomposed into a network of basic functions that together compose the product function. These basic functions are also represented as operations on flows of materials, energies, and signals, which are specified in and retrieved from libraries of basic operations and basic flows, together called a "Functional Basis". Networks or models of basic functions are constructed by specifying for each input-output flow of the product function a chain of operations on flows—basic functions—that step-by-step transform the input flow into an output flow. Models of basic functions, i.e., design representations, are subsequently used to search and compose design solutions.[5] Examples of an overall product function and a model/network of basic functions of an electric screwdriver design are given in Figs. 3.1 and 3.2, respectively.

The model in Fig. 3.2 is a design representation of the intended operation of a-to-be-built artifact, in casu a power screwdriver. It displays part of the mechanism by which the screwdriver would work if it were to be built, i.e., some of its temporally ordered behaviors.[6] Such a partial description of a mechanism thus

---

[5]I here consider 'representation' in a broad sense, which may include models qua diagrams, physical models, drawings, cardboard models, etc.

[6]As we saw in Chap. 2, the concept of 'function' is used with different meanings in engineering design, notably 'purpose', 'effect of behavior', and 'intended behavior'. Product and basic functions in the Functional Basis method refer to 'intended behaviors' (Vermaas 2009; van Eck 2011a).

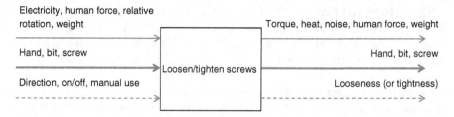

Fig. 3.1 Overall product function of an electric power screwdriver design. *Thin arrows* represent energy flows; *thick arrows* represent material flows, *dashed arrows* represent signal flows (adapted from Stone and Wood 2000, p. 363, Fig. 2)

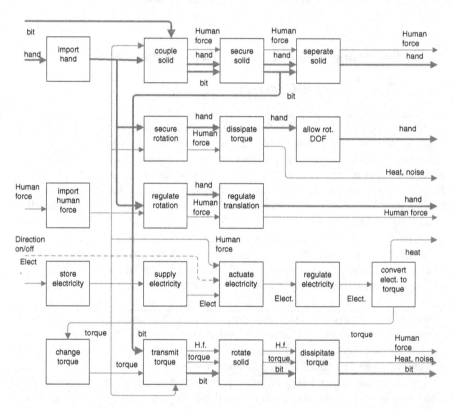

Fig. 3.2 Model of basic functions of an electric power screwdriver design. *Thin arrows* represent energy flows; *thick arrows* represent material flows, *dashed arrows* represent signal flows (adapted from Stone and Wood 2000, p. 364, Fig. 4). Somewhat confusingly the model in Fig. 3.2 is asserted by Stone et al. (1998, 2000) to be a model of a reverse engineered power screwdriver, using the reverse engineering methodology of Otto and Wood (1998), i.e., a model of an extant artifact (see Chaps. 1 and 2). The same model however is also used to illustrate the FB design methodology (Stone and Wood 2000), where it is taken to be a design model of a to be-built-product. Please bear in mind this difference. In this chapter I thus use this model in the design representation sense, i.e., a model of a to-be-built artifact

partially predicts how the screwdriver would work and realize its product function. The model or design representation is *predictive*. And its predictive traction is increased when design solutions for the basic functions are found and configured. When the functional model is aligned with a model or design representation of the (organized) components that are to fulfill the basic functions, a deeper prediction is procured how the screwdriver would work and realize its product function, since more details of its mechanism, i.e., both functions and components, then are known (cf. Craver 2007).[7]

Specifically, such predictive design representations or models provide understanding in *counterfactual* fashion (cf. Glennan 2002, 2011; Casini 2015). Truth-makers of the potential difference making factors listed in the model—temporally ordered behaviors and/or components—are artifacts that as-of-yet still have to be built according to the design representation (and interventions on those artifacts to assess whether the listed components and behaviors indeed function as expected, which is common practice in reverse engineering and redesign (Otto and Wood 2001). So such (counterfactual-supporting) representations are in our design case *not to be assessed in terms of alethic norms*, since the truth makers are still not around. Rather, the goodness of such design representations can be assessed in terms of, amongst others, their plausibility, which, as in our house design-example, relates to the likelihood that the factors listed in the model, indeed, prove to be the correct difference makers with respect to the operation of the intended artifact. One important source for such plausibility assessments is knowledge of past designs and the artifacts constructed in terms of them. Stone and Wood's (2000) 'Functional Basis' (FB) method for designing, again, captures this point nicely. The Stone and Wood lab for designing, for instance, makes use of an automated design tool, called concept generator, in which information on known (technologically and scientifically feasible) components and functions are stored as well known (technologically and scientifically sound) configurations of these components (Bryant et al. 2006). It furthermore makes use of algorithms in order to select the best component configurations for a given design task. Models or design representations of organized components and functions that are constructed in terms of such knowledge base-assisted design tools thus derive their plausibility from knowledge of sound extant designs and artifacts, and technological and/or scientific principles and regularities that apply to them (hence, the more a design is constructed in terms of sound knowledge of extant designs and knowledge, the more prima facie plausibility it has).

The predictive credentials of design models and representations can also be invoked to procure *counterfactual understanding* by specifying and answering a number of "what-if-things-had-been-different questions" (Woodward 2003).

---

[7]Function-component mappings are systematically supported in the Functional Basis method, both in terms of a web-based design repository in which known component-function mappings are archived and a component ontology in which components are archived based on their most commonly ascribed functions (Bryant et al. 2007).

More precisely, by specifying and answering 'what would happen if things would be different questions'. For instance, what would happen when say, some specifics of the conversion of electricity into torque were to be changed, say, when the function 'regulate electricity', or perhaps more precisely 'voltage regulation', were to be fulfilled by a 'voltage regulator' rather than a 'capacitor' (cf. Fig. 3.2). Again, the truth makers of answers to these questions are facts about artifacts that as-of-yet still have to be built (and interventions on them, such as the replacement of components). Still, answers can be given to these questions in the design phase, the plausibility of which, again, derives from sound knowledge of past designs, artifacts that have been build in terms of these designs, and scientific and technological principles governing them. Design models or representations thus assist in counterfactual understanding, and the (goodness of the) understanding they procure can be assessed in terms of their plausibility. Again, alethic norms do not govern such assessments in cases were the artifact has not yet been built/produced (nevertheless such counterfactual understanding may lead to improved designs when *plausible* answers to what-if questions result in the selection of other, better components in the design phase than the ones originally conceived of).

Counterfactual reasoning also looms large in the next case that I consider: reverse engineering and redesign. In this context, counterfactual comparisons that rely on design representations are invoked to show that a redesigned system is expected to function better, relative to a specific functional requirement, than an original system does. I call such predictions with respect to an expected improvement in functional performance of a resigned system *technical advantage* statements. In such contexts, design representations thus fulfill the role of sub serving counterfactual comparisons and the articulation of technical advantage statements.

3.4.2 Prediction and Technical Advantage Statements

Technical advantage statements are common in engineering redesign contexts and aim to provide an answer to the question 'why a redesigned system (RS) is expected to function better than an original system (OS) does'. I illustrate the reasoning underlying such predictive statements in terms of an example of the engineering redesign of an electric wok (cf. Otto and Wood 1998).

In redesign contexts, it is crucial that engineering designers are able to provide an answer to the question 'why a redesigned system (RS) is expected to function better than an original system (OS) does' (cf. Goel and Chandrasekaran 1989; Otto and Wood 1998, 2001; Stone and Wood 2000; Chakrabarti and Bligh 2001; Sen et al. 2011; Sen and Summers 2013). More specifically, 'why RS is expected to fulfill a specific functional requirement (FR) better than OS does'. For instance, in the case of redesigning an electric wok, 'why is it expected that RS-electric wok enables a better uniform heat distribution (FR) than OS-electric wok does' (cf. Otto and Wood 1998).

Answers to such questions, as they are frequently given in engineering redesigning and which I call *technological advantage* (TA) statements, clarify that given a mechanistic organization MO common to RS and OS, and a mode of deployment (MOD) for RS and OS, component C with characteristic C_1 of RS is expected to enable better achievement of FR than component C with characteristic C_2 of OS, or that component C of RS is expected to enable better achievement of FR than component D of OS does. FR is often a role function statement. The prediction thus states that given a MO and MOD, Cs having C_1 enable better performance of a role, relevant to achieving FR, than Cs having C_2 do, or that C performs a role better than D. In the wok case, for instance, that a wok bowl with a certain property, say, a thick bowl, fulfills the role of 'heat conduction' better than a bowl with another property, say a thin bowl, and that therefore the FR of uniform heat distribution is expected to be fulfilled better when the bowl of the wok is thick rather than thin (or thicker than the original bowl) (cf. Otto and Wood 1998).

Developing such TA statements hinges on *counterfactual comparisons* between extant components (with certain properties) and hypothetical ones that differ in parametric details, such as say the thickness of a wok bowl, or hypothetical ones that differ in type, such as say heating coils versus halogen heat lamps (cf. Otto and Wood 1998). These comparisons are aimed to explore which hypothetical component(s) is likely to perform a given technical role better than the extant one. The hypothetical component that is expected to perform optimally is subsequently chosen for redesign. So in order to specify an appropriate redesign, a space of possible options (component variants, parametric and/or adaptive) is explored. This is often done through, inter alia, mathematical modeling of the relevant physical principles and assumptions, and/or prototype building, on the basis of which role performance is compared and a suitable redesign chosen (cf. Goel and Chandrasekaran 1989; Otto and Wood 1998, 2001)

In the context of parametric redesign, specifying a TA statement hinges on comparing a specific type of component with different characteristics (C_1, C_2). For instance, the statement that RS-electric wok has a thicker bowl than OS-electric wok and thereby is expected to perform the role 'conduct heat' in better fashion. In adaptive redesign contexts, giving a TA statement hinges on comparing different types of component-design solutions. For instance, the statement that the halogen heat lamp of RS-electric wok is expected to perform the role 'convert electricity to radiation' better than the heating coil of OS-electric wok does. In both contexts, better performance of these roles is expected to lead to a better achievement of an FR—here, 'uniform heat distribution' (cf. Otto and Wood 1998).

Otto and Wood's (1998, 2001) reverse engineering and redesign method gives an illustration of the above TA statements and the reasoning involved to articulate them. In Otto and Wood's (1998, 2001) method, reverse engineering—mechanistic—explanation drives the subsequent development of a *comparative analysis* between an extant technical system and a to-be-redesigned one with respect to the efficiency with which a specific *role* of a component or sub system is performed relative to an FR.

The details of the reverse engineering phase need not concern us here (see Chap. 2 for these details). Suffice it to say that that this phase is intended to spell out in detail the mechanism(s) by which an extant technical system works, how it is to be operated by users, and to identify components that perform their functional role in suboptimal fashion. This knowledge subsequently drives a redesign phase intended to optimize components that perform their role suboptimal. Optimization can be either parametric or adaptive, depending on the details of the suboptimal role performance (Otto and Wood 1998, 2001).

In this method, the performance of a specific role of a component or sub system of an extant technical system is counterfactually compared with the performance of that role by other, novel components or sub systems—which replace the original component/sub system—in a novel, to-be-redesigned technical system. Counterfactual comparisons of the efficiency with which a specific role is performed between an extant component and hypothetical novel ones are relative to a specific FR, such as the constraint imposed on an electric wok to "deliver a uniform temperature distribution across the bowl" or to "heat and cool quickly" (Otto and Wood 1998, p. 231). That is, TA assessments are relative to specific FRs.

The comparative analysis and subsequent novel component selection either hinges on counterfactual comparisons between extant components (with certain properties) and hypothetical ones that differ in parametric details, such as say the thickness of a wok bowl, or hypothetical ones that differ in type, such as say heating coils versus halogen heat lamps. These comparisons are aimed to explore which hypothetical component(s) performs a given technical role better than the extant one. The hypothetical component that performs optimally is subsequently chosen for redesign. An appropriate redesign is thus specified by exploring and comparing a space of possible options (component variants, parametric and/or adaptive) and selecting the most appropriate one. Technical advantage assessments of redesign options rely in this method heavily on mathematical modeling of the relevant physical principles and assumptions, and/or prototype building.

Specific MOs and MODs that are similar between both the reverse engineered and redesigned system are the backdrop against which such assessments of technical advantages—understood as increased efficiency with which a specific role is fulfilled—are developed. As said, the reverse engineering phase is aimed to detail the MO of the extant technical system that is the focus of subsequent redesign efforts, and also the 'technical habitat' of the extant technical system is explored, i.e., its MOD. So in addition to its MO, also an analysis of user actions ("user functions") with the technical system and interactions between user actions and device functions is carried out (Otto and Wood 1998).[8]

The above is illustrated in Otto and Wood's (1998) method through the redesign of an electric wok. The reverse engineering analysis indicated that an electric wok's

---

[8]Different labels are used in the literature to stress the relevance of user actions in understanding artifact functionality—such as 'user functions' (Otto and Wood 1998); 'mode of deployment' (Chandrasekaran and Josephson 2000), and 'use plan' (Houkes and Vermaas 2010).

FR to "deliver a uniform temperature distribution across the bowl" failed to be fully achieved due to the fact that the electric heating elements of the wok, such as a bimetallic temperature controller, were housed in too narrow a circular channel (Otto and Wood 1998, p. 235). This led to the articulation, in terms of counterfactual comparisons, of a TA statement asserting that both parametric and adaptive modifications were required to components in order to meet this FR in better fashion. Redesign efforts were subsequently directed towards a design with improved functionality of the heating elements, inter alia resulting in a design with a thicker bowl and different shape than in the reverse engineered electric wok, as well as the replacement of the heating coil by a halogen heat lamp.

Specification of FRs, MOs and MODs is crucial in all this, since these provide constraints on which sorts of components can perform certain roles in an efficient manner and which roles are relevant to be performed in the first place. For instance, cooking safely with a wok requires a wok to have handles in order to ensure safe cooking and avoiding burnt hands (Otto and Wood 1998). This FR ('safe cooking') and MOD ('operating the wok by means of its handles') require the wok handles to have a certain shape, mass, etc. to meet these constraints and to adequately fulfill its role of enabling hand-directed manipulation of the wok bowl. Likewise, replacement of the heating coil by a halogen heat lamp which is expected to fulfill the role of "converting electricity to radiation" in a better way than the original wok's heating coil (cf. Otto and Wood 1998, p. 236) is done relative to the FR of 'uniform heat distribution'. Such an FR is required to zoom in on the relevant role(s)—here, electricity-to-radiation conversion—and asses which component does a better job at it. MO specification is vital as well, since components can only fulfill their roles relative to an MO in which they are placed, and given an MO some components do a better job than others. Without, say, appropriate electrical wiring, neither coil and heating lamp can perform their role, and here the halogen heat lamp performs better than the coil does.

The important point is that design representations here fulfill the role of sub serving counterfactual comparisons and the articulation of TA statements. Design models of extant artifacts are used to identify components that function sub optimally and require improving. And models of parametric and/or adaptive (component) redesigns are invoked to counterfactually compare a number of possible redesigns with each other, and with extant technical systems, so as to procure TA statements and select suitable redesigns.

To sum up, what these cases on counterfactual understanding and technical advantage assessments indicate is that it is wrongheaded to attempt to produce alethic statements in terms of design representations and models, since there are still no items in the design phase, i.e., artifacts eventually produced in terms of design representations and models, to utter definite truths (or falsehoods) about. What one rather can and should ask, is whether such models are *adequate* given the design task at hand. Plausibility is a core norm governing such assessments. For instance, is the model 'complete' enough in the sense that all the main functions that are necessary for normal functioning are listed?; are they described with such a level of 'detail' or 'precision' such that components counting as design solutions can be retrieved?; is the 'precision' sufficient to phrase and answer relevant what-if questions; is the model precise enough to identify components that function sub-optimal?; is the model 'plausible' in the sense that its intended operation does not conflict with established technological or scientific principles? For instance, do the redesign models accord with these principles such that indeed suitable redesigns can be selected for manufacturing; etcetera.

Such assessments of models' adequacy is well-established modeling practice (e.g., Levins 1966; Weisberg 2006, 2007). And, often, models tend to exhibit tradeoffs in the sense that one cannot maximize multiple dimensions or desiderata with a single model, indicating the need for multiple models (Levins 1966; Weisberg 2006; Matthewson and Weisberg 2009). For instance, models that are general, being applicable to different systems of the same type, often tend to fare less well on complete and/or precise descriptions of specific token systems, and vice versa. I have argued elsewhere, albeit in different terms, that this also applies to engineering design, thus explaining the engineering practice of using different models of function and functional decomposition (van Eck 2011a, b; see also Chap. 2 for a defense of using multiple models in engineering).

3.5 Conclusion

To understand designing one should know the roles played by design representations in the design process. And some of the central roles of design representations indeed concern 'communication' and 'exploration' (cf. Galle 1999). Yet, as I have been at pains to make clear, such roles should not be conceptualized in terms of communicating and exploring 'truths' about designs. The 'problem of the absent artifact' is a red herring. Rather, I suggest, such roles are better understood in terms of counterfactual understanding and prediction (cf. van Eck 2015 for assessment of other relevant roles). More precisely, counterfactual understanding and prediction are what allow communication and exploration. So, rather than assessing design representations and assertions based on them in terms of alethic norms, a more fruitful approach to spell out the uses and usefulness of design representations is to assess their adequacy with respect to the above mentioned roles. Plausibility is a key constraint on such adequacy assessments.

Analysis of the roles of design representations is important not only for under-standing designing, but also for taking the next step of judging the adequacy of, i.e., testing, design methods. Clarity on the roles design representations fulfill in design methods is a prerequisite for assessing the adequacy of those design representations as they figure in design methods, and thus those methods themselves. Validating or testing design methods is relatively uncharted terrain (van Eck 2015). The results presented in this chapter concerning the roles supported by design representations' for counterfactual understanding and prediction provide a conceptual groundwork for taking further steps of validating those roles. In this chapter I already briefly commented on taking such further steps. As discussed, the adequacy of models can be assessed in terms of desiderata like precision, completeness, abstraction, plausibility, and the like, which are relative to the goals of modelers or designers. Such desiderata can be invoked to assess how successful design representations are in fulfilling their roles, e.g., is the level of detail sufficient to procure the relevant counterfactual understanding, or is the level of detail appropriate to articulate technical advantage statements? In the next chapter we will have a much closer look at the testing of design methods.

References

Bryant, C. R., McAdams, D. A., Stone, R. B., Kurtoglu, T., & Campbell, M. I. (2006). A validation study of an automated concept generator design tool. In *Proceedings of the 2006 ASME International Design Engineering Technical Conferences and Computers and Information in Engineering Conference*, DETC2006-99489: 1–12).

Bryant, C. R., Stone, R. B., Greer, J. L., McAdams, D. A., Kurtoglu, T., & Campbell, M. I. (2007). A function-based component ontology for systems design. In *Proceedings of the International Conference on Engineering Design* (ICED 07): 478.1–12.

Casini, L. (2015). Can interventions rescue Glennan's account of causality? Forthcoming in *The British Journal for the Philosophy of Science*.

Chakrabarti, A., & Bligh, T. P. (2001). A scheme for functional reasoning in conceptual design. *Design Studies, 22*, 493–517.

Chandrasekaran, B., & Josephson, J. R. (2000). Function in device representation. *Engineering with Computers, 16*, 162–177.

Craver, C. F. (2007). *Explaining the brain: Mechanisms and the mosaic unity of neuroscience.* New York: Oxford University Press.

Cross, N. (1992). Research in design thinking. In N. Cross, K. Dorst, & N. Roozenburg (Eds.), *Research in design thinking* (pp. 3–10). Delft University Press.

Galle, P. (1999). Design as intentional action: A conceptual analysis. *Design Studies, 20*, 57–81.

Galle, P. (2008). Candidate worldviews for design theory. *Design Studies, 29*, 267–303.

Glennan, S. (2002). Rethinking mechanistic explanation. *Philosophy of Science, 69*, S342–S353.

Glennan, S. (2011). Singular and general causal relations: A mechanist perspective. In P. Illari, F. Russo, & J. Williamson (Eds.), *Causality in the sciences* (pp. 789–817). Oxford: Oxford University Press.

Goel, A., Chandrasekaran, B. (1989, August). Functional representation of designs and redesign problem solving. In *Proceedings 11th International Joint Conference on Artificial Intelligence (IJCAI-89)* (pp. 1388–1394). Detroit, Michigan August, 1989.

Herbert, D. M. (1993). *Architectural study drawings.* New York: Van Nostrand Reinhold.

Hirtz, J., Stone, R. B., McAdams, D. A., Szykman, S., & Wood, K. L. (2002). A functional basis for engineering design: Reconciling and evolving previous efforts. *Research in Engineering Design, 13*, 65–82.

Houkes, W., & Vermaas, P. E. (2010). *Technical functions: On the use and design of artefacts.* Dordrecht: Springer.

Levins, R. (1966). The strategy of model building in population biology. In E. Sober (Ed.), *Conceptual issues in evolutionary biology* (pp. 18–27). Cambridge: MIT press.

Matthewson, J., & Weisberg, M. (2009). The structure of tradeoffs in model building. *Synthese, 170*, 169–190.

Otto, K. N., & Wood, K. L. (1998). Product evolution: A reverse engineering and redesign methodology. *Research in Engineering Design, 10*, 226–243.

Otto, K. N., & Wood, K. L. (2001). *Product design: Techniques in reverse engineering and new product development*. Upper Saddle River: Prentice Hall.

Pahl, G., & Beitz, W. (1988). *Engineering design: A systematic approach*. Berlin: Springer.

Sen, C., & Summers, J. D. (2013). Identifying requirements for physics-based reasoning on function structure graphs. *AIEDAM, 27*(3), 291–299.

Sen, C., Summers, J. D., & Mocko, G. M. (2011). A protocol to formalize function verbs to support conservation-based model checking. *Journal of Engineering Design, 22*(11–12), 765–788.

Stone, R. B., & Wood, K. L. (2000). Development of a functional basis for design. *Journal of Mechanical Design, 122*, 359–370.

Stone, R. B., Wood, K. L., & Crawford, R. H. (1998). A heuristic method to identify modules from a functional description of a product. *ASME proceedings*, 1–21.

Stone, R. B., Wood, K. L., & Crawford, R. H. (2000). A heuristic method for identifying modules for product architectures. *Design Studies, 21*, 5–31.

Tweedale, M. M. (2004). Future contingents and deflated truth-value gaps. *Noûs, 38*(2), 233–265.

van Eck, D. (2011a). Supporting design knowledge exchange by converting models of functional decomposition. *Journal of Engineering Design, 22*(11–12), 839–858.

van Eck, D. (2011b). Incommensurability and rationality in engineering design: The case of functional decomposition. *Techné: Research in Philosophy and Technology, 15*(2), 118–136.

van Eck, D. (2015). Dissolving the 'problem of the absent artifact': Design representations as means for counterfactual understanding and knowledge generalization. *Design Studies*. Online first, doi:10.1016/j.destud.2015.04.001.

Vermaas, P. E. (2009). The Flexible Meaning of Function in Engineering. In *Proceedings of the 17th International Conference on Engineering Design (ICED 09)*:2.113–124.

Weisberg, M. (2006). Forty years of the 'strategy': Levins on model building and idealization. *Biology and Philosophy, 21*, 623–645.

Weisberg, M. (2007). Three kinds of idealization. *The Journal of Philosophy, 104*(12), 639–659.

Woodward, J. (2003). *Making things happen*. Oxford: Oxford University Press.

Chapter 4
On Testing Engineering Design Methods: Explanation, Reverse Engineering, and Constitutive Relevance

Abstract In this chapter I, draw on philosophical literature on (scientific) explanation to assess the goodness of engineering design methods. I focus this analysis on the engineering design practice of reverse engineering and redesign, and elaborate a constraint drawn from the mechanistic explanation literature to assess the goodness of reverse engineering practices and the content of explanatory and design representations resulting from those practices. This constraint concerns the distinction between *causal* and *constitutive* relevance in mechanisms. I spell out two ways in which constitutive relevance assessments give traction to designing: reverse engineering explanation, and design optimization.

Keywords Testing design methods · Constitutive relevance · Reverse engineering explanation · Design optimization

In the previous chapter we were concerned with the role and goodness of design representations of to-be-built artifacts. Here we address the question what makes explanatory representations of extant artifacts good ones. Together, these analyses give traction on the topic of testing the goodness of design methods, both with respect to the roles of design representations of to-be built-artifacts (Chap. 3) and with respect to the adequacy of representations of extant artifacts (this chapter). Since reverse engineering and redesign are intertwined, the latter topic is clearly relevant for the testing of design methods.

Validation or testing of engineering design methods, apart from their perceived successes in engineering, has received little to none systematic treatment in the philosophical and engineering literature (cf. Vermaas 2014; van Eck 2014). In this chapter I, again, draw on philosophical literature on (scientific) explanation to assess the goodness of engineering design methods. I focus this analysis on the engineering design practice of reverse engineering and redesign, and elaborate a constraint drawn from the mechanistic explanation literature to assess the goodness of reverse engineering practices and the content of explanatory and design representations resulting from those practices. This constraint concerns the distinction between *causal* and *constitutive* relevance in mechanisms. I spell out two ways in which constitutive relevance assessments give traction to designing: reverse engineering explanation, and design optimization. Counterfactual understanding, as in Chap. 3, is a key notion in this analysis, now in fleshing out the relevance of the causal-constitutive distinction. I end by showing how this analysis fits within and extends recent philosophical work on the interplay between engineering design and explanation, indicating the (broader) relevance and promise of connecting philosophy of explanation and philosophy of design.

4.1   Introduction

Two recent and related topics of attention in the philosophy of design concern the (disputed) distinction between science and design (Farrell and Hooker 2012, 2015; Galle and Kroes 2014, 2015), and the testing of design methods (van Eck 2014; Vermaas 2014). Vermaas (2014, p. 47) observed that concern about the scientific status of design by design researchers might me due to the concern that "design research does not live up to the standards of science", since "design research does not yet have the means to test and refute design theories and models". In this chapter I take up the second issue of the testing of design methods (and along the way comment on the prospects and limits of fleshing out the first issue).

There is recent philosophical interest in the connection between engineering design and explanation, both with respect to engineering itself (van Eck 2014, 2015a, b, 2016; Levy 2014; Calcott 2014) and with respect to the interface between engineering and branches of biology (Calcott 2014; Calcott et al. 2015; Braillard 2015; Levy 2014; van Eck 2016). These issues are now starting to get discussed in philosophy of science, particularly those branches dealing with explanation, yet have by and large not been picked up in the philosophy of design literature. I here discuss and further extend this work on explanation and in so doing offer a means to test the engineering design practice of reverse engineering and redesign, as well as the content of explanatory representations resulting from that practice.

In Chap. 3 we took up the related project of elaborating the structure and role of design representations in terms of insights from the philosophical literature on (causal-mechanical) explanation (cf. van Eck 2014, 2015a, b). Here I elaborate a constraint drawn from the mechanistic explanation literature to assess the goodness of reverse engineering practices and the content of explanatory representations as used in reverse engineering and redesign contexts, viz. the distinction between *causal* and *constitutive* relevance in mechanisms (Craver 2007). So whereas in the previous chapter we were concerned with design representations of to-be-built artifacts, the focus here is on representations/models of extant artifacts.

I start with briefly discussing and rehearsing the core tenets of mechanistic explanation in Sect. 4.2. I subsequently elaborate in Sect. 4.3 a key aspect in the construction of mechanistic explanations and assessment of the goodness of such explanations: *constitutive explanatorily relevance* in mechanisms. I discuss this constraint against the backdrop of the *mutual manipulability account* of constitutive relevance in mechanisms (Craver 2007) and show in Sect. 3.4 how this account can be brought to bear on assessing the goodness of reverse engineering practices and resultant explanatory representations. I spell out two ways in which constitutive relevance assessments give traction to designing: reverse engineering explanation, and design optimization. I then show in Sect. 3.5 how this analysis fits within recent philosophical work on the interplay between engineering design and explanation, indicating the (broader) relevance and promise of connecting philosophy of explanation and philosophy of design. One result is that the notion of "evolvability" or modifiability (Calcott 2014), in addition to software engineering, also marks a common core between biology and electro-mechanical design. I end this final section with conclusions.

4.2   Mechanistic Explanation: Explanation by Decomposition

Vermaas (2014) argued that work from the philosophy of the natural sciences, specifically Lakatos' (1978) approach towards falsification and research programs, provides a means to secure a scientific signature for design research and enables the testing of design methods. I applaud Vermaas' (2014) approach to the issue, not only addressing the demarcation issue on 'science versus design', but taking the further step of assessing whether methods or approaches from philosophy of science have traction in the testing of design methods. My contribution in this chapter exemplifies this perspective. Vermaas' proposal is (still) programmatic however since such testing along Lakatosian lines is currently not being carried out in design research. Ultimately, Vermaas offers general guidelines that design researchers may pick up to start the project of the comparative testing of design theories, models, and programs. This is an enormous task and long-term endeavor, for it would require fleshing out in plausible fashion, in the context of design research, all the relevant key concepts of Lakatos' machinery, like theories' hard core, protective belt, associated positive and negative heuristics, empirical content, empirical success, as well as clear comparative measures between competing theories and models.

I rather choose to stay closer to explanatory practices here and focus on recent work from the (mechanistic) explanation literature to elaborate what we might call a 'positive heuristic' that designers may draw on in reverse engineering the workings of complex technical systems and in describing the mechanisms by which such systems (are taken to) work, viz. clearly distinguishing constituent parts of technical mechanisms from causal influences on them.

4.2.1 Mechanistic Explanation

By now, several accounts of mechanistic explanation are on offer in the literature. Although they come in different flavors, there is broad consensus on a number of key features: "All mechanistic explanations begin with (a) the identification of a phenomenon or some phenomena to be explained, (b) proceed by decomposition into the entities and activities relevant to the phenomenon, and (c) give the organization of entities and activities by which they produce the phenomenon." (Illari and Williamson 2012, p. 123). Mechanistic explanations thus explain how mechanisms, i.e., organized collections of entities and activities, produce phenomena (Machamer et al. 2000; Glennan 2005; Bechtel and Abrahamsen 2005; Craver 2007).[1] In the literature on explanation in the life sciences, it is now uncontested that mechanisms play a central role in explaining capacities such as digestion, pattern recognition, or the maintenance of circadian rhythms. The idea is that to explain such capacities, one provides a model, or more generally a description/representation, of the mechanism responsible for that capacity (cf. Chap. 2).

It is clear that mechanism discovery (a, b, and c) is key to the construction of mechanistic explanations (Machamer et al. 2000; Bechtel and Richardson 1993/2010; Craver 2001, 2007; Illari and Williamson 2010). Functional and structural 'decomposition' and subsequent 'localization' of operations/activities on components (Bechtel and Richardson 1993/2010) is probably the most extensively discussed discovery strategy or heuristic (cf. Machamer et al. 2000; Glennan 2005; Craver 2002, 2007). Structural decomposition concerns the process of decomposing a mechanism into its constituent working parts/entities, and functional decomposition gives a model of a mechanisms' constituent operations/activities. Mechanistic explanations are built by aligning these decompositions in terms of localizing mechanisms' operations onto working parts, i.e., by ascribing causal roles to the operations of working parts. These decomposition-localization heuristics are core explanatory business in life sciences like neuroscience, cognitive neuroscience, and parts of biology, where the workings of mechanisms are investigated in terms of a variety of bottom up and top down intervention techniques and experiments, such as brain area stimulation studies and neuroimaging.[2]

---

[1]The precise lingo differs; some speak about 'entities' and 'activities', others 'working parts' and 'operations', yet others 'capacities'. These differences need not concern us here.

[2]Other techniques used in experimental practice and discussed in the literature, concern 'schema instantiation' in which abstract mechanism schemata are made less abstract and applied to particular cases, 'forward-backward chaining' in which gaps in the stages of a mechanism's operation are filled in terms of knowledge of a mechanism's operation in preceding and succeeding stages, respectively (Darden 2002; Darden and Craver 2002), and 'modular subassembly' in which known types of mechanistic modules are assembled to form a hypothetical mechanistic model (Darden 2002). These procedures depend, of course, on mechanistic knowledge procured by earlier functional and structural decompositions and localizations.

Localization is crucial in all this. If done correctly (a non-trivial matter, if anything), one gains knowledge of which parts belong and contribute to the functioning of a mechanisms and how they do so, i.e., which causal or biological role(s) they fulfill in a mechanism. However, neither the conceptual machinery and the experimental practice of decomposition and localization give an unambiguous handle on the issue which component parts and processes are genuine constituents of a mechanism, and which ones are merely causal background conditions or irrelevant parts (Craver 2007). For instance, it is intuitively very clear that the windscreen wipers do not make a (constitutive) difference to the operation of a car engine, whereas the carburetor does. With respect to the mechanism of the car engine, windscreen wipers are simply irrelevant parts. But how to spell out relevance versus irrelevance in clear—cut fashion? Craver's (2007) *mutual manipulability account of constitutive relevance* is devised to handle this problem and spell out when entities' activities are constitutively relevant, i.e., genuine components, of mechanisms rather than causal background conditions or simply irrelevant parts.

To be sure, constituency is crucial to mechanistic explanation. Explanation in terms of mechanisms requires clarity on the (internal) 'make-up' of mechanisms and (external) causal influences on their functioning. Without clarity on what comprises a mechanism, i.e., what its constituents are, in a given explanatory context, that is, what makes up the explanans, explanation becomes vacuous.

4.3   Mutual Manipulability and the Causal-Constitutive Relevance Distinction

4.3.1   Mutual Manipulability

Constitutively relevant factors are individuated by Craver (2007) in terms of mutual manipulability relationships. On Craver's (2007) account, an entity's activity is considered constitutively relevant to the behavior of a mechanism as a whole if that entity's activity is shown to be a spatiotemporal part of the mechanism, and shown to contribute to the behavior of the mechanism as a whole. The latter is crucial for only parts that contribute to a given overall behavior of a mechanism are genuine components. To use an often rehearsed example, the heart's pumping of blood makes a crucial contribution to the circulatory mechanism's behavior of distributing oxygen and nutrients to the body, whereas the noise generated by the heart does not (cf. Cummins 1975; Craver 2001). Evidence for constitutive relevance is taken to be procured if one can change the overall behavior by intervening to change the entity's activity, and if one can change the activity of the entity by intervening to change the overall behavior. Somewhat more formally, a factor is constitutively relevant if two conditionals are met (Craver 2007, CR1, p. 155, and CR2, p. 159):

(CR1) When ϕ is set to the value of $\phi 1$ in an ideal intervention, then ψ takes on the value $f(\phi 1)$

(CR2) When ψ is set to the value of $\psi 1$ in an ideal intervention, then ϕ takes on the value $f(\psi 1)$

These conditionals cover both scenarios in which interventions change the manner in which ψ or ϕ occur, i.e., their value, as well as ones that lead to the occurrence or elimination of ψ or ϕ (cf. Craver 2007, p. 149). In the latter case, ψ or ϕ would take on the value '1' or '0', respectively. So mutual manipulability relations comprise both constitutive relevance, i.e., difference making, relations with respect to the *occurrence* of explananda phenomena, as well as relations concerning the *precise manner in which explananda phenomena occur* or obtain (cf. van Eck 2015a). Note that although the mutual manipulability account is inspired by Woodward's (2003) account of causal explanation, constitutive relevance is a *non-causal* notion (Craver 2007; Couch 2011). Constitutive relevance relationships are always bidirectional—one can in principle always wiggle both overall behavior and component activity by wiggling component or overall behavior, respectively. With causal relationships this is often not the case (exempting cases of feedback). In addition, the relata in constitutive relationships are not logically independent: the tokening of an overall behavior implies the tokening of component activity, and vice versa. Causes and effects in contrast are taken to be logically independent. Finally, constitutive relationships are synchronic: component activities or overall behaviors taking on a particular value are not temporally prior to one another, but happen concurrently. Causes however are by most taken to precede their effects. Since interventions on either components or overall behaviors alone fail to tease causal and constitutive relationships apart, the bidirectional intervention/mutual manipulability constraint is imposed on constitutive relevance assessments (Craver 2007). Mutual manipulability is devised as a general demarcation yardstick for mechanism individuation across sciences dealing with mechanisms.[3]

In my view, Craver (2007) is basically on the right track in his analysis of constitutive relevance. However, following Baumgartner and Gebharter (2015), I take it that mutual manipulability needs to be extended in a significant way in order to make good on its aim of individuating constitutively relevant parts of mechanisms in each and every context. Without this extension, constitutive versus causal relevance cannot always be teased out in plausible fashion in empirical practice. I explain this below.

---

[3]Of course, the interactions between component parts and operations in a mechanism are causal; the relationship between these components parts and processes and a mechanism's overall behavior (the explanandum phenomenon) is constitutive, i.e., non-causal.

4.3.2 Fat-Handedness and Mutual Manipulability Combined

To be sure, mutual manipulability is not uncontroversial; various extensions and criticisms have been given after Craver's (2007) initial formulation (e.g., Couch 2011; Leuridan 2012; Baumgartner and Gebharter 2015; van Eck 2015c).

I side with Baumgartner and Gebharter (2015) that mutual manipulability in itself is not (always) sufficient to establish conclusive evidence for constitutive relationships, but that combined with demonstrating that there are *only common causes* of a mechanism's overall behavior and some constituent, and no surgical causes of a mechanism's overall behavior, this does provide sufficient (abductive) evidence for constitutively relevant difference makers (cf. van Eck and Looren de Jong 2016). Let me explain.

Given the (assumed) non-causal, constitutive relationship between a phenomenon and a mechanistic constituent, an intervention on either the phenomenon or a constituent will ipso facto alter the value of *both* the phenomenon and the constituent (since they occupy the same region of spatial-temporal space and are not related in terms of cause and—temporally later—effect). Such interventions are thus 'fat-handed' (cf. Woodward 2003, 2008; Baumgartner and Gebharter 2015), i.e., all interventions that satisfy mutual manipulability are *common cause* interventions on both the phenomenon and some constituent, altering (the value of) both. Phrased differently, given or assuming constitution, it is not possible to change solely the value of a constituent without altering the value of the phenomenon, and vice versa. So, surgical interventions—which lead only to changes in parts but not in phenomena, or vice versa—should not be not possible. For instance, when wiggling memory formation by engaging a subject in an experimental task would not lead to changes in LTP formation, such an intervention would count as surgical. On the other hand, when such an intervention alters the value of both memory formation and LTP formation—a much more plausible scenario—the intervention counts as fat handed.

However, the problem now becomes that correlations between changes in a phenomenon and some constituent can be explained in terms of their common cause(s), i.e., the intervention(s), rather than putative constitutive relationships. That is, it need not be the case that observed correlations in changes in a phenomenon and some putative constituent are due to constitutive relationships between them; correlations might simply result from the 'fat-handed' nature of the intervention. For example, an intervention that wiggles—effects a change—in both some aspect of Long-Term-Potentiation (LTP) and some aspect of memory formation might suffice to explain the correlated changes in LTP and memory formation due to the 'common cause' nature of the intervention. It seems that there is no further empirical evidence on offer to conclude that constitution grounds the observed correlation:

mutual manipulability via common cause interventions provides no empirical evidence in favor of the existence of constitutive dependencies. Thus, (MM) [mutual manipulability] is not sufficient to account for constitution on evidence-based grounds. (Baumgartner and Gebharter 2015, p. 20).

However, when one combines mutual manipulability with demonstrating that there are *only* common causes of a mechanism's overall behavior and some constituent, and *no* surgical causes of a mechanism' overall behavior that would only alter some aspect of the phenomenon, this does provide sufficient (abductive) evidence for constitutively relevant difference makers (Baumgartner and Gebharter 2015). If there are only common causes, and no surgical causes, the best explanation for this feature is that the relationship between a mechanism' overall behavior and some putative mechanistic component is one of constituency. If there are no causal relationships on offer solely linking an intervention to a change in either a component or phenomenon, whilst interventions that result in changes in *both* component and phenomenon do exist, constitution explains this observation. That is:

constitution provides the best available explanation for systems satisfying both mutual manipulability and fat-handedness (Baumgartner and Gebharter 2015, p. 2)

Importantly, constitution provides a better explanation than the idea of a common cause intervention, i.e., a causal rather than a constitutive relation. If there are no surgical causes/interventions that would enable effecting changes solely in a phenomenon, but not in a putative component, whilst there do exist common cause interventions that effect changes in both, the assumption of constitution explains the correlated changes in phenomenon and putative constituent better than the common cause-notion does, since *constitution also explains the absence of surgical causes* (cf. Baumgartner and Gebharter 2015). So when it is the case that the dependencies between a phenomenon and some constituent cannot be screened off by surgical interventions, constitution offers the best explanation for the observed correlation. For instance, when it is the case that every intervention carried out on some aspect of memory formation changes that aspect of memory formation as well as some aspect of LTP, and there are no interventions that change memory formation but leave LTP unaffected, the best explanation is that LTP is a constituent in the mechanism(s) for memory formation.[4]

---

[4]Some may object that this analysis is vulnerable to counter examples, and hence fares no better—and perhaps worse—than extant criteria for system demarcation advanced in the philosophy of science. Oxygen intake say is required for every cognitive system to function and hence circulatory mechanisms would count as constitutive parts of them. Of course severe interventions on oxygen intake, say suffocating a subject, are fat handed for they shut down the functioning of each putative component as well as the phenomenon targeted for explanation. I feel that such a scenario is outside any sensible request for explanation. More importantly though, the notion that the absence of surgical causes is required for claims on constituency blocks counter-examples such as this one. Craver (2007, pp. 157–158) gives a nice example: interfering with the heart by inhibiting its functioning interferes with word-stem completion, but stimulation of the heart—within certain ranges—does not. Conjoining inhibition and stimulation experiments here suggest that heart function is a relevant causal background condition, not a constituent in word-stem completion

This is of course an example of abductive reasoning, contingent on the current state of play in the relevant sciences. If for a given case only common cause interventions are known and no surgical interventions are available, one has fallible (abductive) evidence for constitution since it explains the absence of surgical causes better than a causal analysis. Yet, this does not rule out in principle that at some point in the future surgical causes might be found. Science is never finished, hence every naturalist analysis is in principle fallible.

With mutual manipulability plus fat handedness, we have solid tools, or so I argue, to test the goodness of aspects of the engineering design practice of reverse engineering and redesign as well as the content of explanatory representations resulting from that practice. This of course concerns the distinction between causal and constitutive relevance. In what follows, the top down and bottom up constraints of mutual manipulability do most work in this testing.

4.4   Testing (Reverse) Engineering Design Methods: Applying Mutual Manipulability

4.4.1   Mechanistic Reverse Engineering Explanation

In engineering, reverse engineering and engineering design go hand in glove (e.g. Otto and Wood 1998, 2001; Stone and Wood 2000). Otto and Wood's (1998, 2001) method for reverse engineering and redesign gives a clear illustration of this interplay. In their method, a reverse engineering phase in which reverse engineering explanations are developed for existing artifacts, precedes and drives a subsequent redesign phase of those artifacts. The goal of the reverse engineering phase is to explain how existing artifacts produce their overall functions in terms of underlying mechanisms, i.e., organized components and sub functions (behaviors) by which overall (behavior) functions are produced. These explanations are subsequently used in the redesign phase to identify components that function sub optimally and to either improve them or replace them by better functioning ones. Otto and Wood (1998, p. 226) relate explanation and redesign as follows: "the intent of this [reverse engineering] process step is to fully understand and represent the current instantiation of a product. Based on the resulting representation and understanding, a product may be evolved [redesigned], either at the subsystem, configuration, component or parametric level".

(Footnote 4 continued)

mechanisms. Furthermore, engaging a subject in a word-completion task does not change the behavior of the heart or other parts belonging to the circulatory condition (except for very 'unusual' conditions, say task execution with a loaded gun pointed at the subjects' head). The point is that in these scenarios surgical interventions are possible, changing either component function or phenomena *yet not both*. This rules out extravagant constituency claims.

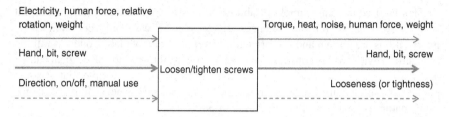

Fig. 4.1 Overall function of an electric power screwdriver. *Thin arrows* represent energy flows; *thick arrows* represent material flows, *dashed arrows* represent signal flows (adapted from Stone and Wood 2000, p. 363, Fig. 2)

In the reverse engineering phase, an artifact is first broken down component-by-component, and hypotheses are formulated concerning the functions of those components. In this method, functions are represented by conversions of flows of materials, energy, and signals. After this analysis, a different reverse engineering analysis commences in which components are removed, one at a time, and the effects are assessed of removing single components on the overall functioning of the artifact. Such single component removals are used to detail the functions of the (removed) components further. The idea behind this latter analysis is to compare the results from the first and second reverse engineering analysis in order to gain potentially more nuanced understanding of the functions of the components of the (reverse engineered) artifact. Using these two reverse engineering analyses, a functional decomposition of the artifact is then constructed in which the functions of the components are specified and interconnected by their input and output flows of materials, energy, and signals (Otto and Wood 2001). Such models represent parts of the mechanisms by which technical systems operate, to wit: causally connected behaviors of components.[5] They are the end results of the reverse engineering phase and are subsequently used to identify sub-optimally functioning components and so drive succeeding redesign phases. Examples of an overall behavior function and behavior functional decomposition of a reverse engineered electric screwdriver are given in Figs. 4.1 and 4.2, respectively.

In the model in Fig. 4.2, temporally organized and interconnected behaviors are described. Components of artifacts are described in Otto and Wood's method in tables, what in engineering are called 'bills of materials', together with a model, called 'exploded view', of the components composing the artifacts. Taken together, these component and behavior functional decomposition models provide representations of mechanisms of artifacts.

After the reverse engineering of a technical artifact, aimed at providing detailed understanding of the mechanism(s) by which it operates, the redesign phase starts

---

[5]To be sure, as mentioned, most have it that the interactions between component parts and processes in mechanisms are causal; the relationships between component parts and processes and overall behaviors of mechanisms are non-causal, constitutive relationships (but see Leuridan 2012 for an alternative construal).

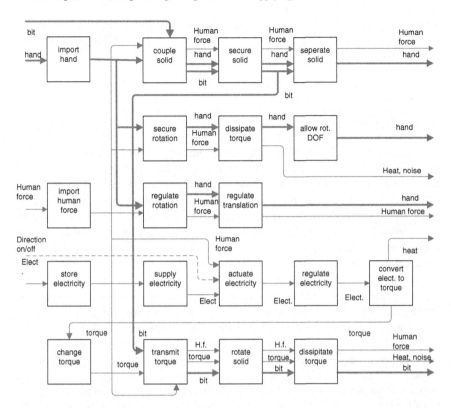

Fig. 4.2 Functional decomposition of an electric power screwdriver. *Thin arrows* represent energy flows; *thick arrows* represent material flows, *dashed arrows* represent signal flows (adapted from Stone and Wood 2000, p. 364, Fig. 4; cf. Stone et al. 1998, 2000)

by identifying components that *function sub-optimally*, and, thereby, cause artifacts to manifest their overall functions in sub-optimal fashion. Redesign efforts are subsequently directed towards designs with improved functionality of these components (Otto and Wood 1998, 2001). Otto and wood (1998) discuss an example of redesigning an electric wok. The (reverse engineered) artifact's desired behavior to "deliver a uniform temperature distribution across the bowl" failed to be achieved due to the fact that the electric heating elements of the wok, such as a bimetallic temperature controller, were housed in too narrow a circular channel (Otto and Wood 1998, p. 235). Redesign efforts were subsequently directed towards a design with improved functionality of the heating elements, inter alia resulting in a design with a thicker bowl and different shape than in the reverse engineered electric wok.[6] In sum, a reverse engineering—mechanistic—explanation of the operation of an

---

[6]This redesign step involves a lot of mathematical modeling, use of physical and technological principles, and/or prototype building (Otto and Wood 1998, 2001). These details need not concern us here.

existing electric wok was used to identify sub optimal functioning components—in this case, electric heating elements—which resulted in modifications to these components.

4.4.2   Testing Case

The model in Fig. 4.2 of a reverse engineered electric screwdriver also gives a clear illustration were things can go wrong in reverse engineering explanation (and mechanism individuation and mechanistic explanation in general): not every component operation represented in Fig. 4.2 is a constituent part of the mechanism by which the electric screwdriver operates. This reverse engineered model is described in terms of a functional modeling language, called Functional Basis, that is taken to only represent device functions, i.e., operations-on-flows carried out by technical artifacts (Stone and Wood 2000; Hirtz et al. 2002; van Eck 2010). With respect to this model, Stone et al. (1998) state that the top chain of functions represents the insertion and removal of the screw bit, that the second represents the fastening of the screw bit, that the third represents the positioning of the screw-driver, and that the fourth and fifth represent the actuation of the device.

However, despite the model and the Functional Basis in general being advertised as describing solely device functions, not every operation-on-flow described in the model in fact represents a device function; quite a few represent operations-on-flows *carried out by users* (van Eck 2010). All the functions of the top function chain and the leftmost function of the second function chain of the power screwdriver exemplify the characterization of user functions given by Hirtz et al. (2002), i.e., operations-on-flows carried out by users. As can be seen in Fig. 4.2, the first function chain is represented in terms of four functions that transform the flows "hand", "bit", and "human force" from input to output. By representing the insertion and removal of the screw bit in terms of a sequence of functions that transform a material "bit" flow, a "human force" flow, and a "hand" flow, the (de)coupling of the screw bit is represented as a sequence of user functions. More specifically, the (de)coupling of the screw bit is represented as realized through human force applied through the hand, i.e., operations-on-flows carried out by a user. This analysis applies as well to the leftmost function "secure rotation" of the second function chain, which represents the manual fastening of the screw bit. In this function chain, the function "secure rotation" transforms a "human force" flow and a "hand" flow, describing that the securing operation is realized by human force applied through the hand.

Now, erroneously interpreting these functions as device functions leads to incorrect understanding of the functioning of the mechanism in question, which in turn is detrimental to redesign and optimization efforts, as well as design knowledge sharing. Mutual manipulability gives a handle on this issue. Although one can envisage bottom up interventions that affect user actions and thereby the overall functioning of the power screwdriver, say, applying too much or too little manual force when driving in screws, the reverse does not (necessarily) hold. Intervening to change the overall functioning of the screwdriver by changing the materials or resistance of the materials in which screws are driven or removed need not have an effect on the hand grip of the user operating the device. The intervention certainly will not have an immediate/synchronous effect on the action of fastening or loosening the screw bit by a user. In other words, there here exist surgical causes/interventions that would enable effecting changes solely in a phenomenon—the driving of screws—without affecting putative components—user actions. User actions are not constitutive parts of the mechanisms of technical systems, here a power screwdriver (but of course they are relevant causal influences on the workings of such systems).

Not only can the conflation of user actions and device functions be ruled out with mutual manipulability. It also can be put to work in teasing apart genuine device functions from (physical) inputs or causal influences on technical systems. In the Functional Basis method for designing, operations-on-flows that represent how input (materials, energy, and signals) enters a technical system also count as device functions (Stone and Wood 2000; Hirtz et al. 2002; cf. Ookubo et al. 2007). We saw above that such functional descriptions may refer to user actions rather than device functions. In other cases such descriptions may refer to input to or causal influences on a technical system, rather than being device functions. Consider again the model of a reverse engineered electric screwdriver in Fig. 4.2. Human force is being modeled as being imported into the screwdriver. This of course is quite sensible, but do such operations-on-flows count as genuine device functions of the screwdriver? On Functional Basis terms they do, but applying mutual manipulability tells a different story. Without the input of human force the screw bit of the screwdriver cannot be fastened/decoupled ("regulate rotation") and the screwdriver hence will not perform its overall function of driving screws. The bottom up condition is hence trivially satisfied. However, intervening on this overall function, again say, by changing the materials or resistance of the materials in which screws are driven or removed will not have an (immediate/synchronous) effect on the human force recruited for fastening or loosening the screw bit by a user.

Not only are some operations-on-flows at the system boundary ruled out as genuine constituents of technical systems. Also some operations-on-flows at the 'center' of the mechanism description fail to conform to mutual manipulability. Consider the two descriptions 'dissipate torque' in the second and fifth function chain. Top down interventions that affect the overall function likely have an effect on the dissipation of torque: increasing the resistance of the materials in which screws are driven or removed impacts the amount of torque that gets dissipated. Yet the reverse, bottom up constraint does not hold. Whether large or small amounts of torque spread out and disappear makes no difference to the functioning of the screwdriver. The operation-on-flow 'dissipate torque' is not a constituent difference maker in the screwdriver mechanism for the driving of screws. (to be sure, torque is relevant for screwdriver functioning, its spreading out however is not!)

Again, clarity on which features comprise a technical system's mechanism and which features are causal inputs to such a mechanism or comprise its "mode of deployment" (Chandrasekaran and Josephson 2000), are crucial for understanding its functioning. And, hence, crucial for redesign purposes and knowledge sharing.

Fat handedness need not do much work in the above case, since the bi-directional constraints of mutual manipulability already sufficed to rule out spurious components. But one can envisage that the fat handedness constraint proves relevant when testing component device functions that prima facie have a constitutive relevance signature. Say, the functions "supply electricity" or "transmit torque" of the power screwdriver. If reverse engineering experimentation would rule out surgical interventions that would only change the overall function of, in this case, the power screwdriver and not these component functions, constitution would best explain the relationship between these component functions and overall function (of course, if such testing were to be carried out this would be the likely result: "supply electricity" and "transmit torque" prima facie do seem constituents of a screwdriver's overall function of driving screws).

4.4.3   The Goodness of Design Representations

The point of course is that good reverse engineering practices and resultant explanatory models or representations highlight bona fide constitutively relevant components and distinguish these from (relevant) causal input, user actions, and irrelevant parts.[7] As alluded to in the case above, the value of making these distinctions lies in their ability to offer sound understanding of the workings of technical systems. We can make this idea precise in terms of a reverse engineering model or design representation's ability of offering adequate *counterfactual understanding*. The model in Fig. 4.2 is a representation of the operation of a technical system, in casu a power screwdriver. It displays part of the mechanism by which the screwdriver works, i.e., some of its temporally ordered behaviors.[8] Such a partial description of a mechanism thus partially explains how the screwdriver works and realizes its product function. I've argued in chapter 3 that an important role of design representations—representations of to-be-built artifacts—is their ability to offer counterfactual understanding in terms of offering answers to *what-if-things-would-have-been-different questions* (van Eck 2015b; cf. Woodward 2003).

---

[7]I use the term representation in a broad sense, which may include models qua diagrams, physical models, drawings, cardboard models, etc.

[8]The concept of 'function' is used with different meanings in engineering design, notably 'purpose', 'effect of behavior', and 'intended behavior'. Product and basic functions in the Functional Basis method refer to 'intended behaviors' (Vermaas 2009; van Eck 2011).

For instance, returning to our screwdriver example, what would happen when say, some specifics of the conversion of electricity into torque were to be changed, say, when the function 'regulate electricity', or perhaps more precisely 'voltage regulation', were to be fulfilled by a 'voltage regulator' rather than a 'capacitor' (cf. Fig. 4.2).[9]

Models that include descriptions of spurious components of mechanisms—be it spuriously identified user actions, causal influences, or irrelevant parts as genuine components—partially fail with respect to this role. Spurious aspects procure incorrect understanding or none at all. For instance, asking how changes in the value of torque dissipation affect the overall function of the screwdriver is an ill-posed question. Torque dissipation is irrelevant for understanding the screw driving mechanism of the artifact, hence no explanatory traction is gained by an inquiry into interventions on its value with respect to screwdriver function.

Consequences of asking the wrong what-if questions with respect to the effects that interventions on user actions and causal inputs have are far more serious. Interventions that change the values of these parameters, of course, often have an effect on overall mechanism function, but is it crucial to know the nature of that effect. Changes in overall device function that result from changes in user actions or causal inputs but are incorrectly taken to result from changes to device functions, gives incorrect understanding of the workings of mechanisms. Misreading changes to an artifacts mode of deployment as changes to its mechanism is nothing short of a category mistake. Redesign/optimization efforts, inter alia, are compromised if these different issues are lumped together, since:

> Giving good explanations is tightly coupled with our ability to manipulate and control the world […] The better we understand the results of various manipulations on some system, the better we can explain how it works. And the better we understand how to control a system by manipulating its parts, the better we can design and build a mechanism with the precise capacities we desire (Calcott 2014, p. 296).

If, however, interventions on component device functions are collapsed with interventions on its mode of deployment we have poor explanation and understanding, and design and manufacture are then the worst for it.

---

[9]Although the truth makers of answers to these questions are facts about artifacts that in a design phase still have to be built (and interventions on them, such as the replacement of components), answers can still be given to these questions in the design phase, the plausibility of which derives from sound knowledge of past designs, artifacts that have been build in terms of these designs, and scientific and technological principles governing them. Design models or representations thus assist in counterfactual understanding, and the understanding they procure in design phases can be assessed in terms of their plausibility. Alethic norms do not govern such assessments in cases were the artifact has not yet been built/produced (nevertheless such counterfactual understanding may lead to improved designs when *plausible* answers to what-if questions result in the selection of other, better components in the design phase than the ones originally conceived of) (van Eck 2015b). See Chap. 3 for more details.

4.5   Outlook and Conclusions

We have seen that import of concepts from the philosophical literature on explanation—here, mutual manipulability—has relevance for the testing of (engineering) design methods (cf. van Eck 2014). This connection also has relevance for the philosophy of explanation. One recent project at the interface of biology and engineering concerns elucidating, and re-characterizing the nature of the relationship(s) between these domains (Calcott 2014; Levy 2014; Calcott et al. 2015). Historically, processes of designing have been likened to biological evolutionary processes (Calcott 2014). Such 'adaptionist' thinking has recently been criticized for providing misleading characterizations of (engineering) designing and, in effect, obscuring import commonalities between biology and (engineering) design (Calcott 2014). One important commonality that has been overlooked concerns the notion of "evolvability" or modifiability that is common to the development of both biological and engineered systems. As Calcott asserts:

> Complex integrated systems, whether evolved or engineered, share structural properties that affect how easily they can be modified to change what they do (Calcott 2014, p. 294)

Evolvable properties refer to features that affect how capacities of systems, engineered and evolved, change over time. Interestingly, although philosophy is only recently picking up on this theme, biologists and engineers alike have been stressing such joint principles governing change for more than a decade (e.g., Csete and Doyle 2002; Kitano 2004; Tomlin and Axelrod 2005). Modularity and robustness are two features that have gotten substantial attention in this context. Calcott (2014) analyzed this common core in the context of biology and software engineering.

The analysis given in this chapter extends this connection to biology and electro-mechanical engineering design. As in biology, evolvability also plays an important role in the context of the reverse engineering and redesign of electro-mechanical systems. Good reverse engineering explanations provide insight into the structure of extant technical systems, making it possible to modify or adapt parts such that optimization of system functionality ensues. Modularity here looms large of course, for this system feature makes it possible to optimize or change parts without affecting other functionalities of the system (in negative fashion). Ease of evolvability or modifiability is thus a desirable feature of technical systems, and good reverse engineering explanations, by highlighting the modular architecture of the functionalities of (genuine) constitutive parts, make it possible to evolve or optimize such systems.

This extension of the connection between biology and engineering, under the rubric of evolvability, is based on this paper's main objective of elucidating the fruitful interplay between philosophy of (scientific) explanation and engineering design, specifically with regard to the testing of engineering design methods. As we saw, the mechanistic concept of constitutive relevance and its assessment in terms of the mechanistic mutual manipulability account, gives means to test the goodness of reverse engineering and redesign practices and the content of explanatory representations resulting from them.

I would like to end this chapter with the suggestion that analyses of relevant interfaces between design and other fields of inquiry, and philosophies thereof, are a more versatile means to spell out what a philosophy of design has to offer and needs to address than analyses solely oriented on scrutinizing the scientific credentials of design. Simon (1969) championed the idea that design and science are relevantly different kinds of endeavors, which has been the status quo ever since. Recently, however, there has been a lively debate between Farrell and Hooker (2012, 2015) on the one hand, who disputed this distinction, and Galle and Kroes (2014, 2015) on the other, who attempted to reinvigorate the position that science and design are distinct, albeit related, kinds of intellectual study. What have such analyses brought us? There is still no commonly agreed yardstick for demarcating science from design that all parties in the debate agree on, if there is such a yardstick to be had in the first place. Differences in method(s), output/product, or aims signal crucial differences to some, yet highlight important commonalities to others. More importantly, little seems to be gained by this 'demarcation debate'. Rather than settling the question whether design is a branch of science or not, a more constructive approach that I hope to have elaborated in this chapter concerns analysis of the roles played by key concepts in the design enterprise, attempts to test them, and assessments of whether methods and/or products from other (scientific) fields or philosophies thereof offer relevant constraints to flesh out such testing attempts. With respect to the latter issue, this paper charted a relevant role for (scientific) mechanistic explanations and mechanistic constitutive relevance assessments. I suspect or at least hope that this is only the beginning. The philosophy of scientific explanation offers a rich source of diverse models of explanation that might prove relevant in the further elucidation and testing of design methods.

References

Baumgartner, M., & Gebharter, A. (2015). Constitutive relevance, mutual manipulability, and fat-handedness. *The British Journal for Philosophy of Science*. Online first, doi:10.1093/bjps/axv003.

Bechtel, W., & Abrahamson, A. (2005). Explanation: A mechanist alternative. *Studies in History and Philosophy of Biological and Biomedical Sciences, 36*, 421–441.

Bechtel, W., & Richardson, R. C. (1993/2010). *Discovering complexity: Decomposition and localization a strategies in scientific research*. MIT Press.

Braillard, P. A. (2015). Prospects and limits of explaining biological systems in engineering terms. In: P. A. Braillard & C. Malaterre (Eds.), *Explanation in biology* (pp. 319–344). Springer.

Calcott, B. (2014). Engineering and evolvability. *Biology and Philosophy, 29*, 293–313.

Calcott, B., levy, A., Siegal, M. L., Soyer, O. S., & Wagner, A. (2015). Engineering and biology: Counsel for a continued relationship. *Biological Theory, 10*, 50–59.

Chandrasekaran, B., & Josephson, J. R. (2000). Function in device representation. *Engineering with Computers, 16*, 162–177.

Couch, M. (2011). Mechanisms and constitutive relevance. *Synthese, 183*, 375–388.

Craver, C. F. (2001). Role functions, mechanisms and hierarchy. *Philosophy of Science, 68*, 53–74.

Craver, C. F. (2002). Interlevel experiments and multilevel mechanisms in the neuroscience of memory. *Philosophy of Science, 69*, S83–S97.

Craver, C. F. (2007). *Explaining the brain: Mechanisms and the mosaic unity of neuroscience.* New York: Oxford University Press.

Csete, M. E., & Doyle, J. C. (2002). Reverse engineering of biological complexity. *Science, 295*, 1664–1669.

Cummins, R. (1975). Functional analysis. *The Journal of Philosophy, 72*, 741–776.

Darden, L. (2002). Strategies for discovering mechanisms: Schema instantiation, modular subassembly, forward/backward chaining. *Philosophy of Science, 69*, S354–S365.

Darden, L., & Craver, C. F. (2002). Strategies in the interfield discovery of the mechanism of protein synthesis. *Studies in the History and Philosophy of the Biological and Biomedical Sciences, 33*, 1–28.

Farrell, R., & Hooker, C. (2012). The Simon-Kroes model of technical artifacts and the distinction between science and design. *Design Studies, 33*, 480–495.

Farrell, R., & Hooker, C. (2015). Designing and sciencing: Response to Galle and Kroes. *Design Studies, 37*, 1–11.

Galle, P., & Kroes, P. (2014). Science and design: Identical twins? *Design Studies, 35*, 201–231.

Galle, P., & Kroes, P. (2015). Science and design revisited. *Design Studies, 37*, 67–72.

Glennan, S. (2005). Modeling mechanisms. *Studies in the History and Philosophy of the Biological and Biomedical Sciences, 36*(2), 375–388.

Hirtz, J., Stone, R. B., McAdams, D. A., Szykman, S., & Wood, K. L. (2002). A functional basis for engineering design: Reconciling and evolving previous efforts. *Research in Engineering Design, 13*, 65–82.

Illari, P., & Williamson, J. (2010). Function and organization: Comparing the mechanisms of protein synthesis and natural selection. *Studies in History and Philosophy of Biological and Biomedical Sciences, 41*, 279–291.

Illari, P., & Williamson, J. (2012). What is a mechanism? Thinking about mechanisms across the sciences. *European Journal for Philosophy of Science, 2*, 119–135.

Kitano, H. (2004). Biological robustness. *Nature, 5*, 826–837.

Lakatos, I. (1978). Falsification and the methodology of scientific research programmes. In: I. Lakatos, J. Worrall, & G. Currie (Eds.), *The methodology of scientific research programmes* (pp. 8–110). Cambridge University Press.

Leuridan, B. (2012). Three problems for the mutual manipulability account of constitutive relevance in mechanisms. *The British Journal for the Philosophy of Science, 63*(2), 399–427.

Levy, A. (2014). Machine-likeness and explanation by decomposition. *Philosopher's imprint, 6*, 1–15.

Machamer, P. K., Darden, L., & Craver, C. F. (2000). Thinking about mechanisms. *Philosophy of Science, 57*, 1–25.

Ookubo, M., Koji, Y., Sasajima, M., Kitamura, Y., & Mizoguchi, R. (2007). Towards interoperability between functional taxonomies using an ontology-based mapping. In *Proceedings of the International Conference on Engineering Design (ICED 07)*, August 28–31, 2007. Paris, France.

Otto, K. N., & Wood, K. L. (1998). Product evolution: A reverse engineering and redesign methodology. *Research in Engineering Design, 10*, 226–243.

Otto, K. N., & Wood, K. L. (2001). *Product design: Techniques in reverse engineering and new product development.* Upper Saddle River: Prentice Hall.

Simon, H. A. (1969). *The sciences of the artificial.* Cambridge: MIT press.

Stone, R. B., & Wood, K. L. (2000). Development of a functional basis for design. *Journal of Mechanical Design, 122*, 359–370.

Stone, R. B., Wood, K. L., & Crawford, R. H. (1998). A heuristic method to identify modules from a functional description of a product. In *Proceedings of 1998 ASME Design Engineering Technical Conferences*, September 13–16, 1998. Atlanta, Georgia, USA.

Stone, R. B., Wood, K. L., & Crawford, R. H. (2000). A heuristic method for identifying modules for product architectures. *Design Studies, 21*, 5–31.

Tomlin, C. J., & Axelrod, J. D. (2005). Understanding biology by reverse engineering the control. *PNAS, 102*(12), 4219–4220.

van Eck, D. (2010). On the conversion of functional models: Bridging differences between functional taxonomies in the modeling of user actions. *Research in Engineering Design, 21*(2), 99–111.

van Eck, D. (2011). Supporting design knowledge exchange by converting models of functional decomposition. *Journal of Engineering Design, 22*(11–12), 839–858.

van Eck, D. (2014). Validating function-based design methods: An explanationist perspective. *Philosophy and Technology*, online first, doi:10.1007/s13347-014-0168-5.

van Eck, D. (2015a). Mechanistic explanation in engineering science. *European Journal for Philosophy of Science, 5*, 349–375.

van Eck, D. (2015b). Dissolving the 'problem of the absent artifact': Design representations as means for counterfactual understanding and knowledge generalization. *Design Studies, 39*, 1–18.

van Eck, D. (2015c). Reconciling ontic and epistemic constraints on mechanistic explanation, epistemically. *Axiomathes, 25*(1), 5–22. doi:10.1007/s10516-014-9243-x.

van Eck, D. (2016, forthcoming). Mechanisms and engineering science. In P. Illari & S. Glennan (Eds.), *Routledge handbook of philosophy and mechanisms*. Routledge.

van Eck, D., & Looren de Jong, H. (2016). *Studies in History and Philosophy of Science, 59*, 11–21.

Vermaas, P. E. (2009). The flexible meaning of function in engineering. In *Proceedings of the 17th International Conference on Engineering Design (ICED 09)*:2.113–124.

Vermaas, P. E. (2014). Design theories, models, and their testing: On the scientific status of design research. In: A. Chakrabarti & L. T. M. Blessing (Eds.), *An anthology of theories and models of design*. Springer.

Woodward, J. (2003). *Making things happen*. Oxford: Oxford University Press.

Woodward, J. (2008). *Invariance, modularity, and all that: Cartwright on causation*. Hartmann: In S.

Printed in the United States
By Bookmasters